Sustainable Energy and Green Finance for a Low-carbon Economy

Jingyan Fu · Artie W. Ng
Editors

Sustainable Energy and Green Finance for a Low-carbon Economy

Perspectives from the Greater Bay Area of China

Springer

Editors
Jingyan Fu
College of Economics, Institute of Resource, Environment and Sustainable Development
Jinan University
Guangzhou, China

Artie W. Ng
School of Professional Education and Executive Development, College of Professional and Continuing Education
Hong Kong Polytechnic University
Kowloon, Hong Kong SAR, China

ISBN 978-3-030-35410-7 ISBN 978-3-030-35411-4 (eBook)
https://doi.org/10.1007/978-3-030-35411-4

This Springer imprint is published by the registered company Springer Nature Switzerland AG
The registered company address is: Gewerbestrasse 11, 6330 Cham, Switzerland

Foreword by Shanjun Li

As the largest energy-consuming country in the world, China has to deal with the mounting challenges related to environmental pollution that is associated with such vast energy consumption. Its energy consumption is estimated to be about 25% of the world's total, whereas its coal consumption is about 50% and CO_2 emission in the range of 30% of the world. The country's urban areas are allegedly going through some of the worst air pollution and traffic congestion problems, which result in deteriorated quality of life and concerns over economic sustainability. As the country develops and implements policies to tackle these challenges, its lessons learnt are relevant for other developing countries mitigating the adverse impacts of climate change.

Developing and especially fast-growing economies have some of the worse air pollution in the world. The health consequences of air pollution can be very high in these areas. Some of the recent studies on China show that PM2.5 from air pollution would have an adverse impact on healthcare expenditures in both the near and medium terms. It is suggested that the yearly savings in national healthcare expenditures as a result of compliance with the World Health Organization's standard of 10 μg/m3 would be close to $42 billion or roughly 7% of China's total health care spending in 2015. In contrast with the conventional perspective that the morbidity impact is modest in relation to the mortality impact, it is found scientifically the morbidity cost of air pollution is estimated more than half of the mortality cost.

Economic incentives and constraints imposed on energy consumption could produce substantial environmental benefits and need to be factored in as an important part of a country's fiscal system. It is well evidenced that there are substantial environmental damages from ineffective energy consumption. Prior studies have already suggested that there is pervasive mispricing of energy across developed and developing countries alike due to ignorance of the external costs generated by fossil fuels. For instance, implementing energy prices embedded with full costs would lower carbon emissions as well as fossil fuel-related air pollution deaths while reducing the expenditures on public health substantially. Policy makers around the world need to reconsider these economic and financial issues carefully when designing public policies for sustainability.

Against these concerns, it is encouraging to see this publication on sustainable energy and green finance contributed by scholars and researchers from various tertiary institutions in Greater Bay Area of China. Such an interdisciplinary approach with integrative thinking to tackle the problems related to environmental sustainability is particularly inspiring. The Institute of Resources, Environment and Sustainable Development (IRESD) of Jinan University established in 2011, as the main research institute in support of this publication, is an exemplary effort by the academics in facilitating collaborative research on sustainable development policies for such a dynamic economic region in Southern China. IRESD provides a research platform for interdisciplinary integration of complementary academic resources. At present, it is the only research platform focused on environmental and resource economic analysis in Guangdong Province. The institute's research team is mainly from the College of Economics at Jinan University, a high-level discipline group that integrates applied economics and theoretical economics.

In this book, the contributing authors also examine issues about the role of green financial instruments, such as green credit and green bonds, which facilitate more public–private partnership and direct investment opportunities into large-scale infrastructure projects with a clear objective to reduce greenhouse gas emissions and to drive more impactful sustainable developments towards a low-carbon economy. These long-term investments into reforming the traditional industries are critical to enable the nurturing and development of an ecosystem made of new green industries, composed of waste-to-energy power generation, clean energy public transport, renewable energy facilities and energy-efficient power plants as well as other environmentally friendly innovation and design, etc. This book reflects the significance of the coordination efforts for environmental governance among Guangdong, Hong Kong and Macao. It stimulates thoughts about the various potentials, pathways and mechanisms of green financing to promote more adoption of sustainable energy and the development of a real economy that is driven by policy and planning towards sustainability and harmony in the Greater Bay Area. This book should provide a useful reference for both academics and practitioners who are keen to understand the grand opportunities and challenges for adopting more sustainable energy and developing a more sustainable economy for the region under incremental collaboration and integration in the light of globalization.

Prof. Shanjun Li
Kenneth L. Robinson Professor
Dyson School of Applied Economics and Management
Cornell SC Johnson College of Business

Co-Director, Cornell Institute for China Economic Research
Cornell University, Ithaca
New York, U.S

Foreword by Jatin Nathwani

I commend the authors on an exemplary effort to bring to the economic policy community a timely contribution to help advance the global agenda for achieving a sustainable energy future. Sustainable development and strategies for an effective transition away from fossil fuels to a low-carbon economy are not only a matter of urgent priority; there is also a compelling need to provide a coherent framework and a rational basis for action. Although this book brings into focus the perspectives from the Greater Bay Area of China, the underlying discussion of the themes has wider appeal and is of relevance for consideration by policy makers globally.

This book by Professor Jingyan Fu and Professor Artie Ng bridges an important gap in our academic literature between theoretical analysis and application to shape and frame specific policies for use in practical contexts through credible frameworks.

We are at the cusp of witnessing a 'sea change' in the public attitudes towards the threat of climate change. The issue is now centre stage at the highest levels of political discourse and the mechanisms of global governance for trade. Beyond policy makers operating in the public sphere, there is increasing recognition of a fundamental global problem of climate stress among individuals, households and the corporate sector from investment bankers to 'small–medium enterprises (SMEs) to large corporate entities with market capitalization greater than $1 billion. Central banks have also become an important voice in shedding light on the question of transition and insurance risks coupled with the risk of stranding of assets.

The book is well organized and in four parts over 15 chapters comprehensively addresses the key aspects that link economics with policy and sustainability:

(i) 'Economic and financial policy for sustainability'
(ii) 'Pathway towards sustainable energy',
(iii) 'Emerging green financing issues' and
(iv) 'Green infrastructure development'.

I observe that the demographic considerations of global population and the movement towards urbanization with the formation of large clusters in 'mega' cities —not only in Asia but also Africa and Latin America—have enormous implications for the achievement of the UN Sustainable Development Goals.

Whether it is stress on energy requirements, water, food or transport, the need for effective mobilization of global capital is evident. This book will remain a valuable addition to help us navigate our way through difficulties and towards positive outcomes through concerted efforts to achieve global environmental sustainability.

Prof. Jatin Nathwani
Ontario Research Chair in Public Policy for Sustainable Energy
Executive Director, Waterloo Institute for Sustainable Energy (WISE)
Faculty of Engineering and Faculty of Environment

Fellow, Balsillie School of International Affairs (BSIA)
University of Waterloo, Waterloo, Ontario, Canada

Preface

It is our hope to develop an edited book to reflect stakeholders' distinct perspectives about sustainable development of the Greater Bay Area of China (GBA), which is positioned to be potentially one of the most economically vibrant regions of the world: a region comparable to the other bay areas of the developed economies that are dynamic, innovative while embracing diversity and, most importantly, sustainability. The Sustainable Development Goals embraced by the United Nations (UN SDGs) emphasize the significance of environmental sustainability which is well correlated with the human health. A polluting environment not only adversely affects the health of GBA's habitants, their communities and the next generation, but also causes climate change as proven scientifically. Scaling up the investments into sustainable and clean energy for a low-carbon economy is widely considered as a prerequisite for environmental sustainability given the implications of greenhouse gas emissions that instigate climate change. Strategic deployment of green finance to accelerate such implementation is a main emphasis of this book.

These challenges to sustainability faced by the GBA can hardly be tackled with a solitary discipline. Sustainable energy and green finance entail cross-disciplinary knowledge ranging from economics, engineering, finance as well as innovative formulation of public policy and business models that enable partnership with the private investors. Exemplary collaboration among multiple stakeholders in the private sector that facilitates timely allocation of resources into sustainable energy from the capital markets is much desired in the next decade before environmental sustainability deteriorates further concomitantly under the ongoing climate change. Further, responsible investments into technological innovation for clean and sustainable energy as well as timely adoption of intelligent infrastructures would be complementary for a rapid transition into sustainable development in the GBA.

Despite the commonalities among the GBA cities with respect to indigeneity, there are inevitable diversities in social and economic development over the region throughout the last century as influenced by a world incessantly undergoing globalization. Such intrinsic heterogeneity is conducive to dynamics in seeking creative solutions for sustainable and clean energy through the process of harmonization towards sustainable development. Complementarity among these cities needs to be leveraged so as to optimize their strengths in facilitating social, environmental and economic sustainability. For instance, Hong Kong and Macau as the Special Administrative Regions have long been positioned as gateways of China for international collaboration and more recently as green financing hubs in relative harmony with the specificities of the global capital markets. Guangdong Province continues to be a vibrant region situated in Southern China being upgraded for innovative, sustainable industry developments that connects seamlessly with the domestic economy.

Engaging diverse stakeholders from various sectors for a healthy and sustainable economy linking the past, current and future generations is never more salient. The efficacy of sustainable development in the GBA would depend greatly on the formulation of policies that are venerated by the regional and local stakeholders who in turn espouse timely and committed execution for sustainability. Such engagement with the stakeholders based on visible, crystalized, long-term common goals is considered critical for the success of implementing sustainable development. As alluded by Thomas Alva Edison, the world-renowned inventor of electric light and power, "*Vision without execution is hallucination*".

The four main sections in this book intend to provide a systematic structure to reveal the interconnected perspectives. Section 1 examines a range of interrelated issues about the formulation of economic and financial policy to stimulate and support sustainable energy developments in the GBA. Section 2 unveils the development of various sustainable energy potentials in the GBA, including case studies about refurbishment of legacy coal-fired power plants, energy-efficient initiatives as well as researches on the growing significance of electric vehicles and smart grid technologies. Section 3 explores the emerging green financing phenomenon and related challenges in developing a sustainable green financing system. Section 4 covers broadened topics of innovative approaches adopted by the various industrial sectors that contribute to sustainability of the GBA's rapid urbanization through complementary efforts, such as fostering the development of green buildings, sustainable travel and effective waste management policy.

Guangzhou, China — Jingyan Fu
Kowloon, Hong Kong SAR, China — Artie W. Ng

Acknowledgements

We are grateful to the editorial team at Springer Nature for their professional support and coordination in the compilation of this book. In particular, we would like to thank Anthony Doyle, Chandra Sekaran and Katrin Petermann.

We are indebted to the contributing authors in this book for their insights, perspectives and timely completion of their work.

We are thankful for the relentless support from our respective institutions as well as the research assistance provided by the Institute of Resource, Environment and Sustainable Development (IRESD) at Jinan University.

Special thanks go to Professor Shanjun Li of Cornell University and Professor Jatin Nathwani of University of Waterloo for their forewords highlighting the interdisciplinarity of this book.

Contents

Emerging Green Financing Issues

Green Infrastructural Development

Editors and Contributors

About the Editors

Dr. Jingyan Fu is Professor at the College of Economics as well as Director of the Institute of Resource, Environment and Sustainable Development (IRESD) at Jinan University. She was Visiting Scholar with the Department of Economics in University of Saskatchewan as well as University of Sherbrooke in Canada between 2005 and 2007. She was also Research Fellow with the Department of Applied Economics and Management in Cornell University in the USA sponsored by the Fulbright Foundation between 2012 and 2013. Her research funded by National Social Science Foundation of China (Ref: 19BJY079) focuses on environmental protection and regulations in relation to sustainable development and growth of the economy in China. She has received the Guangdong Outstanding Achievement Award for her research achievement in environmental regulation and policy

Dr. Artie W. Ng is Principal Lecturer and Deputy Director with the School of Professional Education and Executive Development at The Hong Kong Polytechnic University. His research interest is pertinent to sustainability performance of knowledge-intensive organizations and responsible financing. He has been Visiting Scholar with the Department of Management Sciences at the University of Waterloo, Canada, as well as Visiting Professor with University of Bergamo, Italy, and Jinan University, China. Dr. Ng is currently an International Associate with the Centre for Social and Environmental Accounting Research (CSEAR) at the University of St Andrews, Scotland. He has published with high-impact refereed journals in the area of sustainability, including *Journal of Cleaner Production, Journal of Sustainable Finance and Investment as well as Renewable and Sustainable Energy Reviews*. His scholarly contributions also appear in *Handbook of Green Finance and Encyclopaedia of Sustainability in Higher Education* published by Springer.

Contributors

Entela Benz-Saliasi Department of Finance, School of Business and Management, The Hong Kong University of Science and Technology, Kowloon, Hong Kong SAR, China

Joseph H. L. Chan School of Professional Education and Executive Development, College of Professional and Continuing Education, The Hong Kong Polytechnic University, Kowloon, Hong Kong SAR, China

Theresa Y. M. Chan School of Professional Education and Executive Development, College of Professional and Continuing Education, The Hong Kong Polytechnic University, Kowloon, Hong Kong SAR, China

Jingyan Fu College of Economics, Institute of Resource, Environment and Sustainable Development, Jinan University, Guangzhou, China

Jueqian Gao Institute of Resource, Environment and Sustainable Development, College of Economics, Jinan University, Guangzhou, China

Saisai Guo Industrial and Commercial Bank of China Guangdong Branch, Guangzhou, China

Songdan Nancy Guo School of Professional Education and Executive Development, College of Professional and Continuing Education, The Hong Kong Polytechnic University, Kowloon, Hong Kong SAR, China

Yui-yip Lau Division of Business and Hospitality Management, College of Professional and Continuing Education, The Hong Kong Polytechnic University, Kowloon, Hong Kong SAR, China

Shuyu Li Institute of Resource, Environment and Sustainable Development, College of Economics, Jinan University, Guangzhou, China

Jialin Lin Centre Branch of the People's Bank of China, Zhuhai, China

Junyan Liu College of Finance and Trade, Guangdong Industry Polytechnic, Guangzhou, Guangdong, China

Xuebing Lu China Citibank Guangzhou Branch, Guangzhou, China

Artie W. Ng School of Professional Education and Executive Development, College of Professional and Continuing Education, The Hong Kong Polytechnic University, Kowloon, Hong Kong SAR, China

Christelle Not Earth Sciences Department, Swire Institute for Marine Sciences, The University of Hong Kong, Hong Kong SAR, China

Hao Tan Newcastle Business School, University of Newcastle, Newcastle, Australia

Shuyao Wang College of Economics, Institute of Resource, Environment and Sustainable Development, Jinan University, Guangzhou, China

Menghao Wu Personal Finance Department, Jiangsu Suning Bank Co., Ltd, Nanjing, China

Yang Andrew Wu School of Professional Education and Executive Development, College of Professional and Continuing Education, The Hong Kong Polytechnic University, Kowloon, Hong Kong SAR, China

Tai Ming Wut School of Professional Education and Executive Development, College of Professional and Continuing Education, The Hong Kong Polytechnic University, Kowloon, Hong Kong SAR, China

Economic and Financial Policy for Sustainability

Green Finance Reform and Innovation for Sustainable Development of the Greater Bay Area: Towards an Ecosystem for Sustainability

Jingyan Fu and Artie W. Ng

Abstract The Greater Bay Area of China has reckoned the significance of sustainability in furthering its economic development and growth. This chapter aims to reveal the central planning policy approach that sets the main objectives for an intergenerational sustainable development and growth in one of the most dynamic economic regions of the world. It is envisaged as an opportunity to align with local, regional, and international stakeholders' interests about issues related to health and sustainability, embracing the United Nations Sustainable Development Goals (UN SDGs) and global climate change concerns. In order to reduce the external costs associated with unsustainable developments and to leverage on an emerging green financing system, certain constraints and targets within an ecosystem in alignment with international standards are desirable to enable adoption of effective policies and technological innovation for harmonizing the complementary clusters, societies, and systems for sustainability through timely adoption of renewable and sustainable energy facilities that are beneficial to the next generation of habitants.

Keywords Greater Bay Area · Ecosystem · Sustainability · Sustainable energy · Green finance

1 Introduction

On 18 February 2019, the Central Committee of the Communist Party of China (CPC Central Committee) and the State Council of P.R. China (State Council) officially issued the Outline Development Plan for the Guangdong-Hong Kong-Macao Greater Bay Area (hereinafter referred to as the Development Plan) [3]. As deliberated in the Development Plan [3], "*The Guangdong-Hong Kong-Macao Greater*

J. Fu
College of Economics, Institute of Resource, Environment and Sustainable Development, Jinan University, Guangzhou, China

A. W. Ng (✉)
School of Professional Education and Executive Development, College of Professional & Continuing Education, The Hong Kong Polytechnic University, Kowloon, Hong Kong SAR, China
e-mail: artie.ng@speed-polyu.edu.hk

J. Fu and A. W. Ng (eds.), *Sustainable Energy and Green Finance for a Low-carbon Economy*, https://doi.org/10.1007/978-3-030-35411-4_1

Bay Area (Greater Bay Area) consists of the Hong Kong Special Administrative Region (HKSAR), the Macao Special Administrative Region (Macao SAR) as well as the municipalities of Guangzhou, Shenzhen, Zhuhai, Foshan, Huizhou, Dongguan, Zhongshan, Jiangmen and Zhaoqing in Guangdong Province (hereinafter referred to as "the nine Pearl River Delta (PRD) municipalities"), covering a total area of 56,000 square kilometres with a combined population of approximately 70 million at the end of 2017" (see Table 1). In particular, it recommends ecological development and environmental protection as priorities for developing a quality living area. The Development Plan includes three main aspects for the financial development planning of the Greater Bay Area: constructing an international financial hub, developing a financial industry with unique characteristics, and promoting the connectivity of financial markets.

This initiative is expected to not only reinforce the national policy for sustainable development, but also align with the United Nations Sustainable Development Goals (UN SDGs) [17]. Sustainable development has been found increasingly critical to a healthy living environment and mitigating the external costs in public health instigated by pollution [6, 18]. In terms of a financial industry with unique characteristics, the Development Plan outlines different plans based on green finance for Hong Kong, Guangzhou, Macao, Shenzhen, and other cities.

Table 1 Basic demographics in the Greater Bay Area (*Source* Adapted from Census and Statistics Department of Hong Kong, Statistics and Census Service of Macao, the statistics bureaux of the relevant PRD cities, Hong Kong Trade Development Council, 2018)

Area or city	Size of population (million)	Land area (sq.km)	GDP (USD billion)	Per-capita GDP (USD)
The Greater Bay Area	71.16	56,904	1,642	23,075
Hong Kong	7.48	1,107	363	48,673
Macau	0.67	33	55	82,609
Guangzhou	14.9	7,434	345	23,497
Shenzhen	13.03	1,997	366	28,647
Foshan	7.91	3,798	150	18,992
Dongguan	8.39	2,460	125	14,951
Huizhou	4.83	11,347	62	12,908
Zhongshan	3.31	1,784	55	16,711
Jiangmen	4.60	9,507	44	9,570
Zhuhai	1.89	1,736	44	24,100
Zhaoqing	4.15	14,891	33	8,050

Notes

1. Based on current market prices
2. Calculated with the yearly average exchange rates
3. 2017 figure

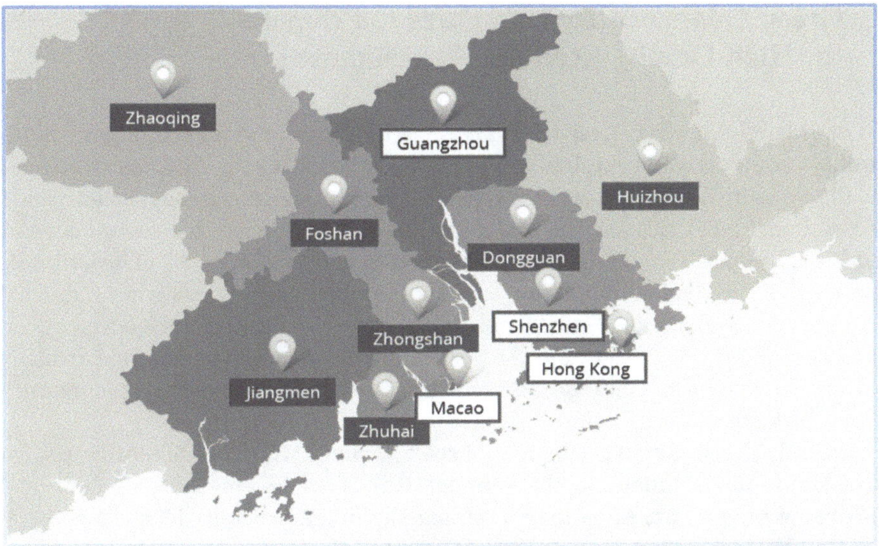

Fig. 1 Core and key node cities in the Greater Bay Area (*Source* Adapted from Constitutional and Mainland Affairs Bureau, Hong Kong SAR Government, https://www.bayarea.gov.hk/en/home/index.html)

Based on the advantages of industrial resources and the geographic characteristics of different areas in the Greater Bay Area, the Development Plan defines Hong Kong as the green financial centre of the Greater Bay Area, Guangzhou as the innovation and experiment zone of green finance reform, and Macao as a complementary green finance platform ([3]; see Fig. 1). In particular, the Hong Kong Special Administrative Region [5] launched its Green Finance Certification Scheme in support of corporate green bond issuance among corporations based in Hong Kong. This initiative is meant to enhance the transparency and accreditation standards of green financial products while strengthening the market confidence towards green finance [5]. Hong Kong Monetary Authority [4] has reinforced this initiative by representing the HKSAR Government in its inaugural green bond offering to refinance public capital projects that enhances environmental sustainability [4].

This approach is intended to enhance a reciprocal opening and connection between the green industry in the Pearl River Delta and the financial market components in Hong Kong and Macao and to encourage the Greater Bay Area to become an international strategic area of green finance development. Such a macro development is considered necessary and feasible to utilize green finance to realize cross-regional collaborative environmental management through the core cities in the Greater Bay Area, namely Guangdong, Hong Kong, Macao, and Shenzhen (see Fig. 1).

2 Green Finance as Driving Force to Promote the High-Quality Economic Development

First, green finance is conceived as an important instrument in enhancing high-quality economic and sustainable development [1, 2]. Green finance can achieve the high-quality development of the real economy through the green transformation of traditional industries and the establishment of a new green low-carbon circular industrial system. Promoting the transformation and upgrading of industry while introducing new factors represents an important approach to resolving industrial land use controversies and further consolidating the foundation for the development of real economy. The Greater Bay Area should rely on the well-established financial industry of Hong Kong and Macao as the driving force and develop green industry with support from the robust real economy in the Pearl River Delta.

Second, green finance can help break through the capital bottleneck of sustainable development in the Greater Bay Area. Ecological protection and restoration projects are often long term, making traditional bank loans inappropriate. Resource allocation through innovative green finance is imperative. Therefore, green finance has emerged with the development of ecological society. Such financing considers environmental factors in financial activities [14]. New institutions have appeared to support the green transformation of traditional industries, complete the foundation of green finance, and ensure the green flow of resources [11].

Third, green finance will help enhance the innovation and collaboration capacity of the environmental governance mechanism in the Greater Bay Area. In environmental governance, goals cannot be achieved depending solely on command and control. Market mechanisms must be allowed to come into play, e.g., to motivate the participation and enthusiasm of polluters to actively reduce their pollution. A green financial system will help introduce social capital into the field of green finance. Subsequently, the efficiency of finance will increase, the financing cost decreases, and the risk disclosure related to environmental issues will be more effective [2, 11].

3 Articulating the Development of Green Finance in the Greater Bay Area

The G20 Summit in 2016 included green finance in the leaders' communiqué for the first time, which resulted in a global blossoming of green finance. Prior to the summit, the People's Bank of China [13], jointly with six other regulatory bodies, published *Guidelines for Establishing the Green Financial System*, which made the launch of green finance development an international effort? The characteristics of the Greater Bay Area, such as being strategic, international, market-oriented, and inclusive, qualify it a well-suited region for the green finance development strategy in China.

3.1 Green Development Policy as Priority in the Economic Planning

As early as 2010, the *Framework Agreement on Hong Kong/Guangdong Cooperation*, the first guideline document on the collaboration between Guangdong and Hong Kong, already included ecological development and environmental protection as the priorities for developing a quality living area. In 2012, the governments of Guangdong, Hong Kong, and Macao jointly released the *Regional Cooperation Plan on Building a Quality Living Area*, which further defined the regional development goals for collaboratively creating a quality living area. The plan specified the direction the collaboration should take, such as environmental protection and changing models of economic development. In March 2016, the State Council's *Guiding Opinions on Deepening Pan-Pearl River Delta (PRD) Regional Cooperation* advocated using driving functions and demonstration effects of Guangzhou and Shenzhen in industry upgrades and green development to develop the Greater Bay Area collaboratively with Hong Kong and Macao. The *2017 Government Work Report* first mentioned planning for the "Greater Bay Area" of Guangdong, Hong Kong, and Macao. In July 2017, the National Development and Reform Commission (NDRC) and the governments of Guangdong, Hong Kong, and Macao signed the *Framework Agreement on Deepening Guangdong-Hong Kong-Macao Cooperation in the Development of the Greater Bay Area*. The agreement, signed in Hong Kong, described a blueprint for the future Greater Bay Area of Guangdong, Hong Kong, and Macao. It also included statements on "ecological priority and green development" and "promoting green, low-carbon production and lifestyle as well as city development and management models." In December 2017, the Central Economic Work Conference specified that the green development of the Greater Bay Area required scientific planning. The development of the Greater Bay Area represents an important collaboration opportunity for the governments of Guangdong, Hong Kong, and Macao [7]. Previous city projects emphasized the economy and finance and typically ignored the social environment, such as the sustainable development of employment opportunities. Future planning will incorporate sustainable development, the economy, the environment, and employment in the development framework of key policies. It will also prioritize green, low-carbon lifestyles, and a suitable living environment in development planning.

Currently, the implementation of green finance policies in the Greater Bay Area remains in a preliminary stage (see Table 2). These green finance policies are not only supporting policies for the top implementation tier of the structural reformation of an ecological society. They are also important supporting measures for the upgrading and transformation of entity industries while providing strategic guidance for green finance and green industries. The development of green finance must rely on the real economy. Prioritizing green and clean industries while limiting the financing provided to industries with high pollution and high energy demand will facilitate the sustainable development of green finance and green industries and bring positive economic benefits.

Table 2 Summary of green finance policies in the Greater Bay Area of Guangdong, Hong Kong, and Macao (*Source* authors)

Timeline	Area	Official document	Content
April 2016	Shenzhen Stock Exchange	*Notice of Shenzhen Stock Exchange on the Pilot of Green Corporate Bond*	Standardizes the procedure for issuing green bonds and identifies the policy support for green bonds, such as "green gateway". Recommends marking green corporate bonds with a "G" label and actively guiding the green bond market in the exchange to support green industries
May 2016	Hong Kong	*Developing Hong Kong as a Regional Green Financial Centre*	Ensures that international capital markets invest heavily in the economic transition towards a green economy and invest assets in green products and projects over the next few years
November 2016	Guangdong	*Implementation Advice on Strengthening Environmental Protection and Financial Integration to Promote Green Development*	Recommends explicitly supporting environmentally friendly enterprises in issuing bonds for direct financing and implementing the issuance of green financial bonds and credit assets' securitization
June 2017	Guangdong	*Interim Measures on the Management of Pu Hui Certified Carbon Emission Reductions*	Mandates formally integrating Pu Hui Certified Emission Reduction (PHCER; the volume of voluntary carbon emission reduction) into the market replenishment mechanism of the carbon emission trading market

(continued)

Table 2 (continued)

Timeline	Area	Official document	Content
June 2017	Shenzhen	*Strategic Framework Agreement of North-South Cooperation to Develop Ecological Civilization and Promote Green Investment and Social Responsibilities*	Mandates actively implementing the country's strategic deployment of green finance and green development. Coordinates cooperation between industries, governments, and regulatory authorities to promote the development of green finance and an ecological society in Shenzhen
August 2017	Zhaoqing	*Zhaoqing Green Finance Proposal*	Proposes increasing the cooperation between government, financial institutes, and enterprises, firmly establishing the concept of green development, actively innovating new development mechanisms of green finance and supporting arrangements, promoting innovation in green financial products and services, and hastening the completion of green financial systems
May 2018	Guangzhou	*Green Enterprise Identification Method in the Green Finance Innovation Experiment Zone of Guangzhou, Guangdong,* and *Green Project Identification Method in the Green Finance Innovation Experiment Zone of Guangzhou, Guangdong*	These identification methods are based on the level of both enterprises and projects. They provide evaluation and identification indicators for green enterprises and projects in different dimensions, such as enterprise management, enterprise environment, technical level of project, and project environment

3.2 Development of Green Finance in the Greater Bay Area

3.2.1 External Comparison with Other World-Class Bay Areas

In 2016, the total economy of the Greater Bay Area of Guangdong, Hong Kong, and Macao was $1.41 trillion USD, which was larger than the economies of the bay areas of Tokyo and San Francisco, respectively. The gross domestic product (GDP) growth (7.9%) was the highest among the top four bay areas in the world. GDP growth in New York, San Francisco, and Tokyo was only 1.5%, 0.7%, and 1.6%, respectively, for the same year. However, compared to other bay areas, gaps remain in industrial structure, GDP per capita, and environmental indicators. The other three major bay areas have nearly completed the upgrade and transformation of industry. In addition, the proportion of tertiary industries remains relatively small in the Greater Bay Area. The air quality indicator of particulate matter (PM) 2.5 was nearly three times higher than that of the New York Bay Area (see Table 3).

Table 3 Comparison of green development between the Greater Bay Area of Guangdong, Hong Kong, and Macao and the other world-class bay areas

Indicator (2016)	Greater Bay Area of Guangdong, Hong Kong, and Macao	New York Bay Area	San Francisco Bay Area	Tokyo Bay Area
Population (ten thousands)	6696	2285	765	3503
GDP (trillions USD)	1.41	1.5	0.8	1.3
GDP per capita (ten thousands USD)	2.11	6.56	10.46	3.54
GDP growth rate (%)	7.9	1.5	0.7	1.6
Proportion of tertiary industries (%)	65	91	83	82
Energy consumption per unit of GDP (kg oil/USD)	0.214	0.129 (U.S.)[a]	0.129 (U.S.)[a]	0.09 (Japan)[a]
PM2.5 (mg/m^3)	30	7	9	12

Source Development Report of the Greater Bay Area of Guangdong, Hong Kong, and Macao (2018), Guangdong Institute of Social Sciences
Note [a]based on an estimate for the country as a whole

3.2.2 Internal Comparison with Other Urban Agglomerations

The premise of achieving sustainable development in the Greater Bay Area is to understand the economic situation and development differences between different cities in the area. To achieve integrated planning for the area, accurate coordination is required. Based on city GDP in 2010–2017, the 11 cities in the area could be divided into four tiers. Hong Kong is in the leading tier. Guangzhou and Shenzhen are in the second tier. The third tier includes Foshan and Dongguan. The other cities, such as Zhongshan, Huizhou, and Zhaoqing, are in the fourth tier. At present, the overall GDP growth of the Greater Bay Area has been steady and doubling. The GDP of the Greater Bay Area rapidly increased to 9.21 trillion yuan (RMB) in 2016 and surpassed 10 trillion yuan (RMB) for the first time in 2017. The overall financial strength of the Greater Bay Area is strong. The region includes Hong Kong, a world-class financial centre, as well as Shenzhen and Guangzhou, where the financial industry has been maturing over time. Therefore, the core cities, such as Guangzhou, Hong Kong, Shenzhen, and Macau, can serve as the main driving force for growth and help marginal cities to promote cooperation on green finance and development.

Development of green finance in Guangzhou

In 2016, Guangzhou became one of the first five national green finance reform innovation demonstration zones approved by the State Council. This position involves important responsibilities to strengthen the major structure of green finance, diversify green financial product systems, perfect the market infrastructure, increase policy support, and develop risk prevention mechanisms for green finance. In response, the Guangzhou municipal government issued a series of documents, such as the *Development Plan of the Regional Financial Centre of Guangzhou (2011–2020)*, *A Three-Year Action Plan (2016–2018) for Developing a Modern Financial Service System in Guangzhou*, and *The Master Plan for Developing the Green Finance Reform Innovation Pilot Zone in Guangzhou, Guangdong.* These documents are all related to policy measures for the development of green finance. According to statistics, in 2016, the amount of loans for ecological protection, green and sustainable energy, emission reduction, and the cyclic economy had reached 50 billion yuan (RMB) in the city of Guangzhou. The number of green enterprises listed on a stock exchange reached 30. The total amount of initial financing was approximately 10 billion yuan (RMB). The unit GDP energy consumption in the area, emissions of carbon dioxide, and emissions of other major pollutants all met the regulating target established by the central government. Table 4 provides a list of power plants under active development in the Guangdong Province.

Further, although Guangzhou does not host many corporate headquarters of financial institutions, it has a large number of regional offices of major financial institutions of a significant scale. Mobilizing financial institutions to implement green finance for sustainable development is an important component of Guangzhou's policy strategy.

Table 4 Major power-generating facilities being developed in Guangdong

Project name	Value (USD million)	Capacity (MW)	Status	Fuel type
Yangjiang Nuclear Power Station, Guangdong	11,903	6,480	Under construction	Nuclear
Taishan Nuclear Power Plant—Phase I, Guangdong	8,700	3,500	Under construction	Nuclear
Geisha Offshore Wind Farm, Guangdong	725	198	Under construction	Wind
Shenzhen East Waste-To-Energy Plant, Guangdong	65.4	168	At planning stage	Biomass
Taishan Nuclear Power Plant—Phase II, Guangdong	N.A.	3,500	At planning stage	Nuclear

Source Fitch Solutions, Infrastructure and Project Finance, China, 28 February 2019

Development of green finance in Hong Kong

Green finance in Hong Kong has experienced two key developments. The first was the decision to have the Special Administrative Region (SAR) government and issuers regulated by public organizations issue green benchmark bonds to motivate more market players to participate [16]. The second was the establishment of a certification plan for green finance projects and securities to attract new issuers and investors to the city [16]. Currently, Hong Kong is using its status as an international financial centre to promote green finance to the mainstream. The number of public and private institutions issuing green bonds in Hong Kong is increasing. These institutions include MTR Corporation, Castle Peak Power Station, Link REIT, Town gas, and China Development Bank. The issued bonds have a variety of currencies and periods. Hong Kong launched the Green Finance Certification Plan in 2018. The city is also preparing to initiate a green sovereign bond issuance plan of $12.8 billion USD, which will set the world's record for the largest amount of green sovereign bonds. Hong Kong is actively working with the upsurge in the development of green finance and striving to play an important role in the regional development of green finance. For example, Hong Kong is assisting enterprises and financial institutions from Mainland China to issue green bonds overseas and to attract investments from Hong Kong and international sources. Through this communication mechanism, it helps international investors enter the bond market in Mainland China. Utilizing the advantage of an offshore renminbi (RMB) market, it can promote the development of green bonds denominated in RMB. Hong Kong is also endeavouring to establish an international green bond index and a platform for funds and wealth management products.

Development of green finance in Shenzhen

In 2018, Shenzhen became one of the cities in the second group of nine green finance reform innovation demonstration zones approved by the State Council. The development of green finance has been showing good progress. Shenzhen's commercial banks have been developing increasingly diversified forms of green finance. The Industrial Bank has invested in the corporate bonds of Dongjiang Environmental Protection, which specializes in waste management. The Bank of China, Shenzhen Branch, has realized asset securitization for the beneficial right of photovoltaics. Financial institutions in Shenzhen fully support ecological development and industrial upgrades to promote green, low-carbon, and cyclic economic growth. They are constantly seeking to unite credit resources with the needs of green development. Meanwhile, data from the Shenzhen Banking Regulatory Bureau indicates that the balance of green finance-related loans in Shenzhen has been increasing at a remarkable rate. In the first half of 2017, the loan balance of energy conservation and environmental protection projects and services in the banking sector in Shenzhen was 100.4 billion yuan, 8.37% higher than the growth rate of other loans during the same period. Loans for clean energy and green transportation projects accounted for a relatively high percentage. Loans for energy-saving and environmental protection services and green construction projects also increased rapidly. On 13 December 2017, Shenzhen announced its accession to the International Network of Financial Centres for Sustainability (FC4S). Joining the FC4S will further enhance Shenzhen's international visibility in green finance.

Development of green finance in Macao

Macao has a small-scale economy, a large foreign dependence, and a monotonous industrial structure. It must change its development model and promote diversification in the economic structure. At present, the development of a green economy has become the consensus of all sectors of Macao. The SAR government has committed to implementing all aspects of environmental protection and improving the level of environmental protection to "build a low-carbon Macao and share a green life." The development of a green economy in Macao has three advantages. First, the industrial structure relies heavily on the service industry. Macao's economic structure has been focused on the tertiary industry. With less investment, low consumption, low pollution, and high efficiency, the development of a green economy offers natural industrial structure advantages. It is reasonable for Macao to become the engine of green economic development. Second, the green development index is high. The green development index primarily includes three categories: the green degree of economic growth, the carrying capacity of resources and environment, and the support of government policy. For Macao, the modern service industry is the main area of green economic growth. Its current environment of low resource and energy consumption is the foundation. The SAR government's confidence and measures to develop a low-carbon economy provide fundamental assurance. The combination of these three factors positions Macao's green development index among the highest in Asia. Third, Macao residents have a strong understanding of the green economy. In addition, the Macao International Environmental Cooperation Forum and the Asia-Australia Forum, hosted by Macao, both focus on low-carbon, environmental protection, and the development of a green economy. These efforts indicate that Macao has paid substantial attention to and been deeply involved in the articulation of a green economy aiming to enhance Macao's international status in environmental protection.

4 Exploration and Practice of National Green Finance Reform Demonstration Zone

4.1 Features of Guangzhou Green Gold Reform Pilot Area

In August 2016, the Central Leading Team for "Comprehensively Deepening Reform" approved the *Guidelines for Establishing the Green Financial System,* which was a milestone for an increasingly developed top-level green finance system in China. The development of green finance requires both top-down design and bottom-up grassroots innovation and exploration. On 23 June 2017, the People's Bank of China and seven other state departments issued the *Master Plan for Developing the Green Finance Reform Innovation Pilot Zone in Guangzhou, Guangdong*, which specifically identified Guangzhou's Hued District as the core area for developing a green finance reform innovation pilot zone in Guangzhou. Among the first group of green finance demonstration zones, areas in five provinces (regions), including Zhejiang, Jiangxi, Guangdong, Huizhou, and Xinjiang, were selected to develop green finance reform innovation pilot zones with different specialities and features. The system mechanism was investigated to search for a replicable experience.

4.2 The Advantages

Among the five provinces (regions), the Guangzhou Green Gold Reform Pilot Area is the most economically advanced demonstration site. As a pioneer in reform and opening-up, the market economic system is relatively mature in Guangdong. The focus of the reform is to investigate a new development model that integrates green financial reform and economic growth.

The Guangzhou Green Gold Reform Pilot Area is the only green gold reform zone with a pilot carbon-trading project, which has unique advantages in the coordinated development of carbon emission trading and green finance. It also has policy advantages in the development of carbon finance and carbon sinks. So far, the innovative carbon finance practice in Guangdong has been in the forefront of the country's seven pilot carbon markets. By the end of 2017, cumulative transactions in the Guangdong carbon market exceeded 1.5 billion yuan (RMB), accounting for 31.7% of the total transactions of all carbon-trading pilots in China. It became China's first pilot carbon market with a spot transaction amount over 1.5 billion yuan (RMB). Guangdong, Hong Kong, and Macao are expected to construct a green financial system and develop a regional green financial centre.

The Guangzhou Green Gold Reform Pilot Area and the Greater Bay Area of Guangdong, Hong Kong, and Macao both represent strategic decisions at the national level. The two strategies complement and promote one another, forming the basic development characteristics of the Guangzhou Green Gold Reform Pilot Area. Within the Greater Bay Area, Hong Kong and Macao have the advantages of being international financial centres. Foshan, Dongguan, and other cities have advantages in manufacturing. The complementary development among cities in the Greater Bay Area provides important support for the development of the Guangzhou Green Gold Reform Pilot Area.

5 Green Development in Guangzhou City Through an Innovative Ecosystem for Sustainable Economic Growth

5.1 Promoting High-Quality Economic Development Through Environmental Protection

To ensure high-quality economic development, it is imperative for the governments involved to develop a set of transparent evaluation criteria for green developments in GBA. In comparison with traditional projects, green projects have certain public attributes. Their key features include energy saving, emission reduction, and a concept of green development. For instance, as of the end of March 2018, 100 electric buses had been put into use in the green finance reform pilot area in Guangzhou. The expected annual reduction of fossil fuels is 1397 tons, that of nitrogen oxide emission is 3492 tons, and that of carbon dioxide emission is 5624 tons. The effect of emission reduction is equivalent to planting 22,000 trees. The introduction of electric buses effectively reduces the exhaust emission of buses in the pilot area and improves air quality.

Being consistent with the policy orientation to serve the real economy, the Guangzhou Green Finance Reform Demonstration Zone has compiled the *Green Enterprise Identification Standard in the Green Finance Innovation Pilot Zone of Guangzhou, Guangdong,* and *Green Project Identification Standard in the Green Finance Innovation Pilot Zone of Guangzhou.* These documents define the standards used to identify green enterprises and projects in multiple dimensions, including enterprise and environment, project technical level, and project and environment. They also define the boundaries for green enterprises and projects and identify over 1200 green enterprises and projects.

Green development has been promoting industry upgrades. Green finance projects encourage entities to save energy and reduce emissions. Their major direction is to support the development of companies that provide energy-saving services. The orientation of investment explicitly supports green finance and industrial development. For example, Guangdong Efficiency Power Plant has achieved reductions in energy consumption and emissions. It also has entirely changed the traditional production and development structures of a utility company by replacing fossil fuels and hydro with renewable energy. The transformation exhibits prominent characteristics of an industry chain upgrade.

5.2 Implementation of Green Development with Innovative Government Support

Some specific measures taken by the Guangzhou City Government to support green development are highlighted as follows:

i. The government directly releases policies and implementation details to promote green development. For example, to promote the development of the Guangzhou Green Finance Reform Pilot Zone, Guangzhou was the first to publish detailed guidelines for the green finance pilot zone on 11 July 2017. On the same day, its Hued District released detailed guidelines for the green finance pilot zone, namely the "1 + 4" matching policy to support green finance and green development.
ii. Government departments rely on green funds to support green projects and innovate green financial market instruments based on market demand. For example, relying on the Energy Efficiency Power Plant Fund from the Asian Development Bank, Guangdong Energy Conservation Centre offers a series of policy measures, including innovative collateral, energy-saving awards and incentives, low-cost financing, long-term loans, and government undertaking of external risks and project management costs. They thoroughly consider the characteristics of specific enterprises (including third-party energy-saving companies, state-owned companies, and private companies) and projects and make substantial contributions to the implementation of the Guangdong Efficiency Power Plant project.
iii. A focus on supporting the green development of small- and medium-sized enterprises and creating new models of green finance. In the progress of developing a green financial reform pilot zone, Hued District established a platform for risk

prevention and control of green finance. The district registered all of its more than 30,000 enterprises and invented an identification procedure for green enterprises and projects. More than 1200 green enterprises and projects were identified. The platform's administrators not only pay attention to the largest energy consumers in the district but also support a number of small- to medium-sized enterprises. Risk control and green certification services are provided to these enterprises, thus enhancing these enterprises within the parameters of Guangzhou's green development practices.

5.3 Adoption of Market Instruments to Motivate Enterprises to Participate in Green Development

In its green development practice, Guangzhou has comprehensively used various marketization instruments, created innovative mechanisms of green development, and enriched the meaning of green development. In this manner, green development in Guangzhou has been significantly promoted. More specifically, Guangzhou's approach has been as follows:

i. Fully utilize the mechanisms of pricing and supply and demand. Establish the concept of compensation for use through market tools, such as the trading of environment rights and carbon. Encourage enterprises to conserve energy, protect the environment, and achieve green development.
ii. Fully utilize the effects of the credit interest rate. Motivate enterprises to convert their green development projects to market projects through interest subsidies and financial incentives. In its green development practice, Guangzhou has successfully converted a number of green development projects to market projects, as represented by the Efficiency Power Plant project. The various effective approaches, including interest discounts, simplified procedures, innovation guarantees, and financial awards, have helped enterprises pass the breakeven point and successfully implement green development projects.

An illustration of an integral ecosystem for sustainable development facilitated by an array of complementary renewable and sustainable energy facilities is depicted in Fig. 2.

Fig. 2 Illustration of an ecosystem for renewable and sustainable energy (*Source* Waterloo Institute of Sustainable Energy, University of Waterloo with permission)

6 Direction for Further Investigation in the Future: Innovation Through Green Finance in the Greater Bay Area

6.1 The Dual Attributes of Green Finance as Both Financial and Public Instruments

Because ecological and environmental protection has a strong public dimension, it is difficult to meet the financial demand for ecological and environmental restoration by relying solely on government support. In this case, how to use the market to provide adequate financing for ecological environmental projects has become a new question in public finance. Green finance involves the entry of private capital into the field of public products. It has both financial and public characteristics. The financial characteristics refer to the demand for rewards for achieving sustainability without pursuing maximum profits. The lower cost of financing makes such rewards possible. Finance has two characteristics with respect to the efficiency of using funds. The first is that finance is an exchange of value across time periods. The second is that finance can connect supply and demand across regions. Green finance connects the demand and supply of ecological environmental resources through the cross-period and cross-regional characteristics of financial means and achieves the market value of ecological environmental resources and the optimal allocation of resources. Compared to traditional finance, green finance also has characteristics of public products, which include the characteristics of asymmetric information, maturity mismatch, and external benefits. Therefore, investigation of financing models is required in the context of policy, finance, and the combination of the two, combined with the specific characteristics of a project. First, for public products, government financial support should be provided. Second, for quasi-public products and profitable public products, joint financing involving government financial support and market financial means could come into play, which in turn enable innovation in the design of green financial products in alignment with a sustainable and healthy living environment. Under such an amiable environment, business and service sectors that require immense R&D as well as technological innovation can be nurtured and expanded.

6.2 Issues with Information Asymmetry and Initiatives to Enhance the Measurement of the Green Standard

Because green projects are characterized by long project terms and long return cycles, financing difficulty is a major obstacle to their development. Green finance is an effective tool to provide financial support for green projects. Green bonds offer a series of advantages, such as long periods and low costs, and are one of the most efficient green financial products. The total amount of China's green bonds issued in 2017 was $36.4 billion USD. Except for coal, certain large hydropower projects, and other projects that do not meet international standards for green bonds, the total amount of issuance of green bonds certified by both international and Chinese standards was $22.5 billion USD. The issuance of green bonds certified according to the Chinese standard alone was $13.9 billion USD. China's total amount of green bond issuance ranked second in the world in 2017. Certain green bonds certified by the Chinese green bond standard are not recognized internationally because of the difference between Chinese and international green bond standards. In addition to these differences between Chinese and international standards, there are multiple Chinese standards. First, the People's Bank of China, NRDC, and the stock exchanges have three different green bond standards. Second, the green bond standards are inconsistent with green credit standards. Assets identified as green credits do not necessarily meet the green bond standard after securitization. In addition, the current evaluation standards simply distinguish green or non-green without any intermediate range. There is no measurement of the degree of greenness. A measurement method is much desired with which one can evaluate the "green value" of the economic activities. Third-party assurance of actual environmental performance of green-financed projects would be an opportunity to deal with the institutional constraints that largely focus on the planning stage of environmental impact assessments in the country [8, p. 12].

6.3 Clarifying the Boundary Between Government and Market for a Long-Term Collaborative Mechanism

How to effectively reduce the cost of green finance to develop green industry is the key to the formation of a long-term mechanism for green industry. While the environmental and social benefits of green development are clear, the economic benefits are difficult to perceive in the short term. Sustained investment of green financial resources is required. The early development of green industry is difficult to achieve depending on market forces alone. Government intervention and promotion are required. However, with limited government financial resources, it is necessary to consider questions such as how to use the leverage of government funds, make green projects more attractive to financial institutions, encourage more social investment in green industries, and develop a mechanism for the withdrawal of government intervention. Green finance is currently primarily invested in individual enterprises, such as sewage treatment and waste incineration projects. A relatively successful case that crosses the boundary between government and business is the Fifth Thermal Power Plant in Guangzhou's Hued District. The project investment was sourced from enterprise and bank loans, whereby loans represent 80% of the total project financing. To receive construction funding and resolve the risk concerns of the bank credit support, the project uses cost to determine the price for waste treatment and provide qualified collateral for project financing. The government has developed a financial balancing plan. The waste treatment fee was set at 148 yuan (RMB) per ton. The project will have an 8% annual profit rate. Meanwhile, government purchase contracts have been signed. The company uses the power generation profit to determine the future rights of return as collateral for bank credits. Based on these facts, is there an ideal linkage point for green finance, city management, and land development? Can we further integrate land and real-estate appreciation to improve regional environment and promote financial innovation on a larger scale?

7 The Way Forward: Augmenting Green Finance Innovation for High-Quality Development

In the future, green financial innovation must respond to the following questions in terms of system and practice. First, how can market-based private capital participate more in green finance with a long-term mechanism be established so as to optimize the government's involvement? Second, how can the external benefits of green finance be integrated into the application and evaluation of projects, so that green finance can be advocated to promote the development of municipal infrastructure? Third, how can we develop a pluralistic co-governance model that encourages stakeholders' and public participation at large?

7.1 *Determining the Industry Standards, Constraints, and Targets for the Development of Sustainable Energy Infrastructure and Green Industry*

Developing green standards is a fundamental issue for green investment and the development of green finance. First, a set of acceptable green finance standards need to be established. The required measures include the formation of green product and service standards, green project identification standards, green credit evaluation standards, and green finance statistics standards that can be widely applied in Guangdong, Hong Kong, and Macao. Second, based on green industry standards, enterprises can be rated based on their degree of environmental sustainability. Such rating will eliminate the intermediate state between green and non-green ones. Second, policies should be derived to incentivize various degrees of "greenness". For example, outstandingly rated green enterprises would receive corresponding tax or industrial land use benefits. Support in the form of loans would also be provided to prioritize the credit demands of these enterprises. Third, the standards for green industry must be studied adopting a holistic approach. Green industry standards should be expanded into a set of integrative green standards applicable to green energy, green industries, green campuses, green enterprises, and green supply chains. Industrial upgrades using the specialties of green finance should be promoted. In particular, a clear objective is to drastically reduce greenhouse gas emission resulting from industries in alignment with the global objective towards zero emission from the power-generating facilities and transportation vehicles that fundamentally empower the overall logistics and supply chain of green industry. Policy makers need to formulate mechanisms to seek multitude views of local and regional stakeholders in developing the pertinent policies for green developments and adoption of renewable energy given the diversity of stakeholders in GBA [10, 12]. In particular, the policy makers can adopt the "backcasting" as a planning tool to determine the targeted timeline of sustainable goals while resolving the hurdles to be overcome over the period of sustainable development prior to attaining such goals [9, 15].

7.2 *Encouraging Innovation in Green Finance Models that Are Sustainable*

A relevant example is the experience of the Asian Development Bank (ADB)'s Guangdong International Loan Project. This project started in 2002 and is the first green credit project in China. The financing arrangement of this project has been changed to extend the repayment period from three to eight years. This change was made due to the fact that the alternative associated energy projects possess three important characteristics: a long-term investment cycle, a relatively large scale, and manageable risk. When ADB loans were first implemented (more than 10 years ago), alternative energy projects were not yet treated as part of the strategic industries in

China. Another reason is the time demanded by energy improvement and transformation. While such projects are not profitable during the period of equipment replacement, the lender requires the enterprises to repay their loans based on the free cash flows generated using their energy savings. Such innovative loan approach enables the external environmental benefit that the environmental enterprises provide to society to moderate the repayment period. This mechanism can extend the time required to resolve the maturity mismatch of green finance.

7.3 Design of Scalable Green Finance Mechanisms

Taking the characteristics of the external benefits of green industry into consideration is essential for industry upgrades high-quality economic development while advocating green city qualities. It is advantageous to embrace environmental ecological values supported by a feasible financing plan for infrastructure development. That is, because of the expected environmental improvement and the potential appreciation of assets of improved sustainability qualities, it is justifiable to introduce green finance as well as investment opportunities that enable the implementation of scalable environmental improvement projects. The motivation for enterprises to participate in green projects largely depends on the extent of whether the government can formulate policies that help safeguard their expected returns while containing the risks involved; thereby, the features of green finance that mitigate the external costs of environmental degradation can be realized.

References

1. Bouake S, Nguyen DK (2019) Corporate social responsibility, ethics and sustainable prosperity. World Scientific Press
2. Financial Services Development Council (2016) Hong Kong as a Regional Green Finance Hub
3. Greater Bay Area (2019) https://www.bayarea.gov.hk/en/outline/plan.html (accessed 27 July 2019)
4. HKMA (2019) HKSAR Government's Inaugural Green Bond Offering. Available at: https://www.hkma.gov.hk/eng/key-information/press-releases/2019/20190522-3.shtml (accessed 2 Aug 2019)
5. HKSAR (2019) Gov't Backs Green Financing. Available at: https://www.news.gov.hk/eng/2019/03/20190325/20190325_171653_647.html (accessed 2 Aug 2019)
6. Landrigan P et al (2017) Lancet commission on pollution and health. The Lancet, Published online 19 Oct 2017, http://doi.org/10.1016/S0140-6736(17)32345-0
7. Lau S (2019) China's Guangdong-Hong Kong-Macao Greater Bay Area: a new development opportunity for Hong Kong. Public Adm Policy Asia-Pacific J 22(1):8–14
8. Lo CWH, Tang SY (2014) Institutions, regulatory styles, society and environmental governance in China. Routledge, UK
9. Ng A (2009) Backcasting performance of the emerging renewable energy sector in China. J Technol Manag China 4(1):53–66
10. Ng A, Nathwani J (2010) Sustainable energy policy for Asia: mitigating systemic hurdles in a highly dense city. Renew Sustain Energy Rev 14(3):1118–1123

11. Ng A (2018) From sustainability accounting to a green financing system: institutional legitimacy and market heterogeneity in a global financial centre. J Clean Prod 195:585–592
12. Ng A, Nathwani J (2018) Sustainable energy infrastructure for Asia: policy framework for responsible financing and investment. In: Bhattacharyya SC (ed) Routledge handbook of energy in Asia. Routledge, UK, pp 284–295
13. People's Bank of China (2016) Guidelines for establishing the green financial system, jointly issued by the People's Bank of China, Ministry of Finance, National Development and Reform Commission, Ministry of Environmental Protection, China Banking Regulatory Commission, China Securities Regulatory Commission and the China Insurance Regulatory Commission. http://www.pbc.gov.cn/english/130721/3133045/index.html [accessed 1 Dec 2018]
14. Price Waterhouse Coopers or PWC (2013) Exploring green finance incentives in China
15. Robinson JB (1982) Energy back casting A proposed method of policy analysis. Energy Policy 10(4):337–344
16. Securities and Futures Commission or SFC (2018) Strategic framework for green finance.https://www.sfc.hk/web/EN/files/ER/PDF/SFCs%20Strategic%20Framework%20for%20Green%20Finance%20-%20Final%20Report%20(21%20Sept%202018....pdf [accessed 8 May 2019]
17. The United Nations (2017) 17 goals to transform our world. http://www.un.org/sustainabledevelopment/globalpartnerships/ (accessed 5 Sept 2017)
18. Yuen PP, Ng AW (2019) Healthcare financing and sustainability. In: Leal Filho W et al (ed) Good health and well-being. Encyclopaedia of the UN Sustainable Development Goals, Springer, Nature Switzerland, pp 1–10

Climate Disclosure and Climate Risk for Asian Companies

Entela Benz-Saliasi

Abstract Climate risk is now recognised to be significant to the businesses and society overall. A fast-changing environment with the new regulation, policy changes, technological shifts and consumer behaviour is expected to put significant pressure on companies. Companies are starting to realise that business-as-usual scenario is no longer valid. Nonetheless, few company leaders seem to grasp the idea of what climate risk means, let alone internalise these risk and carry stress test scenarios. The card of downplaying these risks as too uncertain or distant in future is no longer credible. Company executives need to take measure to create a resilient and financially sustainable business. This paper discussed the relevance of climate risk for companies. While the topic is increasingly catching attention, little work is done documenting and analysing the economic and financial impact of climate change on a country, sector and firm level. This chapter tackles that issue on two levels. First, it analyse the climate disclosure and climate risk for Asian-based companies. Second, we look at country-specific climate risk and its relationship with financial performance. It shows that climate risk disclosure is still in very infant stage in Asia despite the need for much more. The author observes an overall negative relationship between climate risk (and physical risk) and company valuations captured by P/E and P/B ratios. This is an essential result for asset managers, and asset owns for valuations and portfolio allocations. Once considered a hidden and extreme risk, and climate risk, it should be considered a systematic risk and therefore demand a risk premium.

Keywords Climate risk · Climate disclosure · Physical climate risk · Transitional risk · Emerging market · Price to earning ratio · Price to book ratio · Company financial performance

---

E. Benz-Saliasi (✉)
Department of Finance, School of Business and Management, The Hong Kong University of Science and Technology, Kowloon, Hong Kong SAR, China
e-mail: entelabenz@ust.hk

J. Fu and A. W. Ng (eds.), *Sustainable Energy and Green Finance for a Low-carbon Economy*, https://doi.org/10.1007/978-3-030-35411-4_2

1 Introduction

Climate risk is now recognised to be a significant risk to the businesses and society overall. According to MSCI 2018 report on climate change, economic losses are projected to reach at least 14.37% of the country GDP for a high-warming scenario (+6%), and that number excludes equally large climate-related costs such as migration, diseases and social conflicts. Companies are under pressure to report on how climate change affects their operations and financial performance. A more transparent disclosure will be beneficial to the investment community looking to transit to a low-carbon economy. Investors are increasingly expecting companies to report on climate-related risks with as much as rigorous methodologies and clarity as any other financial risk. Climate risk is particularly important for long-term investors. Reducing exposure to carbon heavy sectors does not necessarily reduces performance as shown in Andersson et al. paper [1].

Meanwhile, company executives have barely paid any minimum level of attention to the climate issues. A recent report from McKinsey [16] states, they either fail to understand climate risk or believe that is too distant to effects today's valuations. Their un-action might be fatal for companies operating in certain sectors such as energy, utility or materials. KPMG survey on Corporate Responsibility [14] states that 72% of large and mid-cap companies do not acknowledge the financial risk of climate change in their annual financial reports. And on the minority that acknowledges such risk, only 4% provides some data on the business value-at-risk. The one that reports the data is typically companies from countries where such disclosure is either mandatory or encouraged from the government, regulatory or stock exchange institutions.

In order to measure the climate risks, the industry (alongside IFC and World Bank) has developed some necessary general measures. The very basic one relies again on carbon emission disclosure. The problem with this measure is that it is very narrowly defined and it is backward looking. Ceres [9] uses a climate measure that captures both the textual context and the length of the disclosure. While not perfect, it has already been used by a few academic researches with mixed success.

Meanwhile, academic research[1] on the impact of climate risk is minimal due to the lack of market focus and meaningful data. The lack of market focus is associated with the market misperception of climate risk as a long-term risk. In reality, these risks span from short- to mid- to long-term horizon. Hence, there is an urgency to educate the market, the companies and the investors that these risks will hit them faster than expected. As such, the need for proper analysis and business adaption to climate change is strongly pressing ahead as discussed in Litterman et al. [15]. Recent studies of Hong et al. [11], Jong et al. [13] as well as Cicero [7], Deutsche Bank [10] and Caldecott et al. [8] emphasise that both physical risk and transition risk have a material impact on company level. Schroders Investment Management [17] estimates that companies would have to spend at least 4% of their market value to ensure their asset against physical climate risks. In 2017, the Task Force on Climate-Related

[1] https://www.cfapubs.org/doi/pdf/10.2469/faj.v72.n3.2.

Financial Disclosure[2] released reports and called for better climate risk reporting, climate risk management and stress test scenarios. Given the regulatory pressure and increased market awareness, rating agencies such as Moody and Standard and Poor's are planning to incorporate (or are partially incorporating) such risks into credit ratings.

In the following pages, we aim to provide the investors with valuable insights on climate risk measures and the relationship of climate risk with the company's financial health and finally explore the relevance of specific dimensions of climate risk to company's financial and profitability ratios. By shedding light on the relation of climate risk to company profitability ratios, i.e. ratios that are widely used by investor to screen companies, we hope to decode and simplify the climate risk with the purpose of conveying the clear message of the danger of mispricing climate risk and ultimately motivate the investors to shift their capital from irresponsible and short-sighted businesses to resilient and sustainable ones.

We show that there is a clear trade-off between climate risk and company performance. From a portfolio allocation perspective, investors should be mindful of reducing exposure to countries that have high climate risk. Our intra-region analysis provides a heat map of country exposure to climate risk and its components specifically.

2 Defining Climate Risk

Climate risk is the risk of our planet, society and business face due to climate change, and given the increasing number of evidence-based research, we can, with a certain degree of confidence, attribute that to global warming. Climate risk has in general four components: first, the physical risk of being exposed to extreme weather events as well as climate change; second, the transition risk related to regulatory changes to mitigate climate change; third, the technological advancement of renewables and energy efficiency increase the risk to existing industries relying on fossil fuels; and fourth, it is the social risk due to changes in consumer trends and behaviour.

While climate risk can be decomposed into four categories, we will only focus on the two most relevant categories such as physical and transitional risk. Each of these risks has a significant financial impact on company assets, operations and valuations overall.

Physical risks are related to both climate and extreme weather events such as heat waves, storms, floods, fire, droughts, sea level rise. This once thought to be rare events is happening at not only a much higher frequency but also an increasing intensity. The result is considerable economic and societal damages. Naturally, although not at the desired speed, the market is starting to realise that these impacts are not a one-off,

[2] https://www.fsb-tcfd.org/wp-content/uploads/2017/06/FINAL-TCFD-Technical-Supplement-062917.pdf.

but have profound implications in our ecosystem. As such, companies will have to adapt or face extinction.

Transition risk arises from the transition from high-carbon to low-carbon economy. They are associated with policy, liability, technological with in-depth market and social implications. The additional societal risk while recognised as significant is not included in the calculation given its difficulty and complexity.

A simple analysis of the country to country basis reveals that some countries fare better in physical risk, but worse in transition risk. For example, the MSCI 2018 study reveals that Europe is in physical risk, but not facing a high cost on transition risk.

In order to find out some relationship between climate risk and financial performance, some key performance indicators (hereinafter KPIs) are necessary for quantifications. Given the lack of comprehensive disclosure on climate risk, we need to combine two different data sources: Bloomberg and CDP. The KPIs for climate risk are taken from the CDP website. The Bloomberg data is used for measuring physical risk, while CDP data for transition risk.

A significant number of companies have released specific indicators that are used to measure physical risk and transition risk, respectively.

2.1 Measuring Physical Risk

Physical risk is calculated using BlackRock model[3] and ESG Environmental Disclosure Sector in Bloomberg terminal, namely the quantity of total GHG (or carbon dioxide if GHG not available) emission, total water consumption and total waste disposal.

$$\text{Physical risk} = \text{function (GHG emission, water consumption, waste disposal)}$$

The level of physical risk elicited by each element is determined by its magnitude; for example, if the total GHG emission is 2440.20 metric tons (2.4402×10^3), the physical risk level merely due to GHG emission should be 3. Analogously, physical risk level attributed to water consumption and waste disposal is measured with

[3]Blackrock: Adapting Portfolios to Climate Change, September [5].

• Resource efficiency: Companies that generate more sales with less carbon, water and waste are deploying resources more efficiently. Companies that recycle are rewarded with a higher score while those contributing to landfills are penalised.

• Climate risks: Estimating risks to companies, ranging from the effects of possible carbon taxes to the impact of extreme weather events on labour productivity, then estimates temperature-induced income shocks. It captures how firms perceive their exposure by counting the absolute number and change in disclosed climate-related risks.

• Climate opportunities: Finally, identify potential winners by tracking filed green patents and disclosed climate opportunities, and the annual change on these metrics. This is meant to capture corporate shifts towards alternative energy and innovations such as cleaner chemicals, new wastewater treatments and energy storage.

the same principle. For example, in our sample Li Ning Company Ltd, a Chinese company that makes sports goods, reports the following KPIs:

i. Total GHG emission of $2440.2 = 2.4402 \times 10^3$
ii. Total waste of $337.1 = 3.37 \times 10^2$
iii. Total water consumption of $66540 = 6.6540 \times 10^4$,

The physical risk is therefore equal to $= 3+2 + 4 = 9$. In our analysis, we will exclude companies that have not reported on all three items; i.e. we require full disclosure. The KPIs are therefore all in absolute levels and are not normalised for company size or sales.

2.2 *Measuring Transition Risk*

Transition risk to a low-carbon and resilient climate future is dependent on specific industrial scenarios as well as regulations and policies of the country where the company locates. More severe climate impacts combined with a less successful implementation of the climate disclosure and regulatory framework will likely result in a significantly large and devastating financial risk for companies. In this study, transition risk is calculated based on CDP scoring and the category of industry. CDP scoring is an indicator of a company's performance addressing climate, more specifically environmental issues. The scoring[4] is sector specific, hence more intuitive to interpret climate risk per sector.

Based on the degree of disclosure and the level of discussion of climate-related risk within the company level, the CDP has developed an intuitive approach to presenting scores. The score that highlights a company's progress towards leadership uses a four-step approach:

i. Disclosure: The completeness of the company's response to the CDP questionnaire;
ii. Awareness: The extent to which the company has assessed climate issues, risks and impacts about its business;
iii. Management: To what level has the company implemented actions, policies and strategies to address climate issues; and
iv. Leadership: The steps that a company has taken which represent best practice in the field of climate management (Fig. 1).

The logic of scoring is simple. For example, each company has to satisfy at least 80% of the criteria for the lowest level and a minimum number of the requirement in each category, i.e. disclosure to progress to the next level that is awareness and

[4]It is nonetheless important to clarify that the score does not reflect any technological and societal risks that are equally important, but understandably difficult to quantify and analyse.

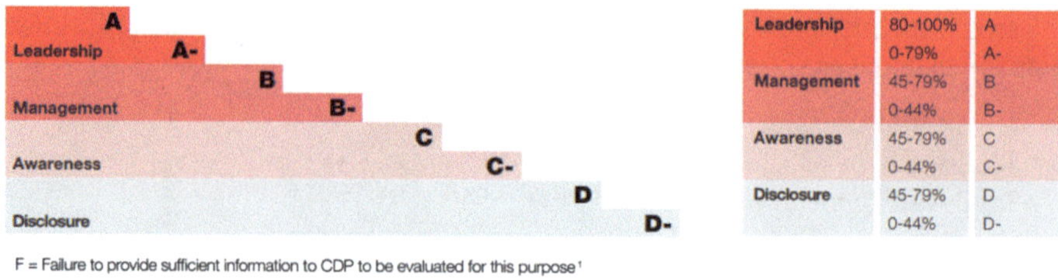

Fig. 1 CDP scoring (reproduced from CDP website) CDP (A-leadership; B-management; C-awareness; D-disclosure; F-failure to provide sufficient information for evaluation)

management. Only a score higher than 80% will lend them to the leadership class. For more, refer to CDP Scoring Introduction 2018.[5]

However, numeric CDP scores for companies requested by CDP questionnaires are not transparent on the CDP website, because only nominal grades are given in the reports for some companies. Therefore, there is a need to translate nominal grades into numeric scores as shown in Appendix 1.

Larger transition risk could incur if the disclosure and regulation framework are not strong enough, meaning a higher CDP score indicates a lower transition risk. Based on BlackRock model, a transition risk is a numeric number ranging from 0.5 to 3. Given the negative relationship between transition risk and CDP score, we calculate transition risk with the following formula:

$$\text{Transition Risk} = 3 - (3 - 0.5) * \text{numeric CDP score}\,\%.$$

Total climate risk is defined as:

$$\text{Climate risk} = \text{Physical risk} + \text{Transition risk}$$

Having defined both physical and transition risks as well as determined the relation between the two, our ultimate objective is to study the extent of climate risk disclosure in Asia. To answer the question, we will look at 350 Asian companies that reported to CDP during 2017–2018. However, before dwelling into the Asia region, there is a need to understand how climate disclosure and climate risk appear in other regions. To set the scene, we consider two groups: emerging market companies and US companies that report to CDP and have data on Bloomberg.

[5]https://b8f65cb373b1b7b15feb-c70d8ead6ced550b4d987d7c03fcdd1d.ssl.cf3.rackcdn.com/cms/guidance_docs/pdfs/000/000/233/original/Scoring-Introduction.pdf?1479494696.

3 Climate Risk Disclosure in Emerging Markets

Given that climate risk is a new topic in the market and the difficulty associated with it, it is normal to expect a low level of awareness and therefore low disclosure not just in Asia, but developed as well as developing countries. To draw a comparison for Asian companies, we study:

i. Companies that sit on the top 50 and bottom 50 MSCI Emerging Market ETF Index as of June 2018;
ii. One hundred and thirty-six US public companies that are among the top global 500 companies which have responded to the CDP questionnaire in 2013. While the US data of 2013 is not directly compatible with the 2017 MSCI data, the year mismatch helps in analysing the level and the speed of climate awareness, climate adaption and climate-related disclosure of emerging countries and Asia in particular versus the USA.

Among the top 50 companies in the MSCI EM ETF Index, only 44 companies have responded to the CDP questionnaire. Among these 44 countries, 3 are from Hong Kong, 8 are from South Korea, 10 are from China, 2 are from Taiwan, 5 are from South Africa, 5 are from Brazil, 4 are from India, 2 are from Mexico, 3 are from Russia, 1 is from Thailand and 1 is from Malaysia (see left side of Fig. 2). Complete environmental KPI disclosure on Bloomberg is only 36.64%.

Among the bottom 50 MSCI EM ETF companies, only 43 companies have responded to the CDP questionnaire. Among these 43 countries, 2 are from Taiwan, 6 are from South Korea, 4 are from Turkey, 1 is from the Czech Republic, 3 are from the Philippines, 2 are from Thailand, 2 are from Mexico, 2 are from the United Arab Emirates, 8 are from Malaysia, 1 is from Egypt, 1 is from South Africa, 1 is from Greece, 5 are from India and 5 are from Indonesia (see right side of Fig. 2). Complete environmental disclosure on Bloomberg: 18.6%.

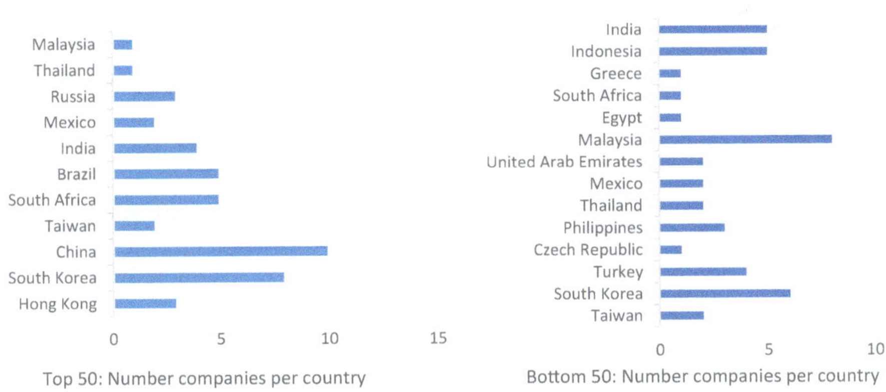

Fig. 2 Distribution of MSCI EM ETF top 50 and bottom 50 companies reporting to CDP

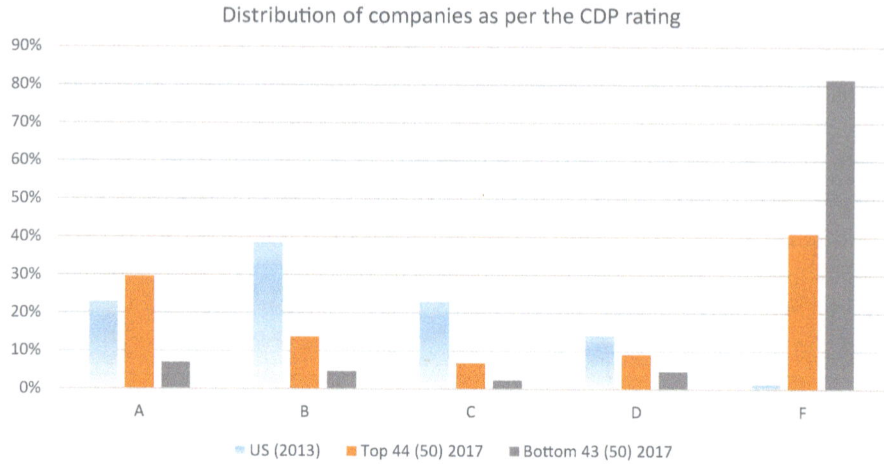

Fig. 3 CDP rating for 136 US companies versus top 50 and bottom 50 MSCI emerging market companies

Looking at the dispersion of companies that occupy the top versus bottom 50 of the MSCI EM Index, a couple of observations emerge from Fig. 2. The market is more disperse for the lower bottom than the top. China is the country with companies that report the most to CDP, while Malaysia has the highest number of companies that while part of the MSCI EM ETF does not report to CDP. South Korean companies are almost equally represented.

Our next point of interest is to investigate the extent to which the emerging market companies are addressing climate risks as per the CDP questionnaire. Figure 3 plots top 50 versus bottom 50 emerging market companies that are part of the MSCI emerging market ETF. Also, we compare their numeric disclosure[6] to 136 US companies reporting to CDP in 2013[7]. The CDP coding stands for A-leadership, B-management, C-awareness, D-disclosure and F-failure on climate-related risks. The desired result would be to have more companies rated at A-level and only a few at D- or F-level.

The results are striking. Most of the emerging market companies are at the F-level, meaning failure to disclose enough information on climate risk that is due to either lack of understanding or lack of awareness of climate risk, or combination of the two. Second, companies seating at the MSCI EM ETF bottom 50 have a deficient climate disclosure relative to the MSCI EM ETF top 50 companies. Put it differently in 2018, 81% of bottom 50 fails to disclose sufficient information relative to only 40% for the top 50 companies and 2% for US companies. Put it differently, even after a gap of 5 years, Asian companies part of the MSCI EM Index and respective leader in their countries are far behind in acknowledging and measuring climate risk.

[6]Fortunately, numeric CDP scores are released on CDP website for all these 136 US companies.

[7]https://data.cdp.anet/Climate-Change/2013-Global-500-Emissions-and-Response-Status/marp-zazk.

Regarding distribution, it is interesting to see that MSCI EM ETF top 50 companies oscillate between leadership and failure to provide sufficient information (30 vs. 40% tails) exhibiting a U shape distribution. This indicates a large dispersion even within the top-rated companies, with some clear leaders in the space and existing opportunities to profit from. In comparison, the 136 US companies exhibit a quasi-normal distribution with positive skewness. The majority of US companies will be concentrated in the body with sufficient disclosure, but not many leaders in the industry. Given the increasing market attention to climate risk-related disclosure, we firmly expect the skewness to be even stronger with more US companies concentrating on the top two rankings (A-level and B-level).

With the growing importance of the topic, we expect more companies to populate the A/B and C area indicating high levels of leadership/management and awareness of climate risk.

4 Climate Risk and Financial Performance

After looking at climate risk disclosure, we further explore the relationship between climate risk and firm performance. We hypothesise that firms with high climate risk should have a low firm valuation or higher beta as per Trinks et al. [20] it is nonetheless very important to keep in mind that climate risk as its many shades are most likely not priced in yet as per Litterman et al. [15]. We are, however, focusing only on correlation that per se does not imply causation. Like the ESG discussion raised by Krueger (2015), this is similarly a chicken-and-egg debate. It is logical to expect firms with high valuations to have overall good risk management and most likely rich resources to report and address climate risk, which in turn will bid up future valuations. The question is when does the kickback happen? When will the market learn to read and price climate risk? When will the more stringent regulation, such as fines and adoption of cleaner technology, eat into the cash flows?

Put it differently, what has so far been of second-order risk and proxy by traditional risk management frameworks will very soon become a primary risk that needs to be priced in. Given the limited and partial data on climate risk, it is challenging to evaluate the impact of climate on valuations. We can, however, test the possibility of a relationship between company valuation and climate risks.

Two financial KPIs are selected to indicate company's valuation, i.e. business performance: price to earnings P/E ratio and price-to-book P/B ratio. The price–earning ratio (P/E ratio) values a company, measuring its current share price relative to its per-share earnings. The price-to-book ratio (P/B ratio) is calculated by dividing price per share by book value per share, and a low P/B ratio could mean that the stock is undervalued. The P/E and P/B ratios are well documented in the financial literature to capture a market anomaly starting with Basu [3, 4]. Empirical results have found that companies with low P/E generate a higher return even after controlling for higher risk. Value investors consider this company undervalued, hence cheap, and recommend buying them betting on a mean reversion pattern. However, this not always the case as low P/E might indicate low growth and hidden risk. And, this is what we are also betting on; climate risk is a hidden but potentially significant risk.[8] Consequently, we should observe a negative relationship between a firm with high climate risk and low P/E ratios.

To capture the relationship, we use a simple but efficient measure such as R-value, that is the Pearson correlation coefficient measuring the strength of the linear relationship between climate risk and financial KPIs. The sign indicated the direction of the relationship with the absolute value and captured the strength.

To test the hypothesis, we refer again to the top 50/bottom 50 MSCI EM ETF companies and US companies. However, our sample size shrinks further from 87 (43 bottom + 44 top companies) to 67 companies as these are the only ones that have responded to CDP and have climate disclosure data on Bloomberg.

As expected, Fig. 4 confirms our expectation; that is, climate risk and firm profitability are negatively correlated. The result is robust across the spectrum, in the USA as well in Asia, for MSCI EM ETF top 50 or bottom 50 companies. Companies that rank high on climate risk also have the lowest profitability ratios such as P/E and P/B. Without getting lost into the causality, it is essential for investors to recognise that low P/E companies face higher climate risk, a significant risk largely ignored and unpriced. These risks are not short-term. Hence, low P/E does not imply an underpriced company, but rather a persistent abnormality that it most likely becoming more pronounced.

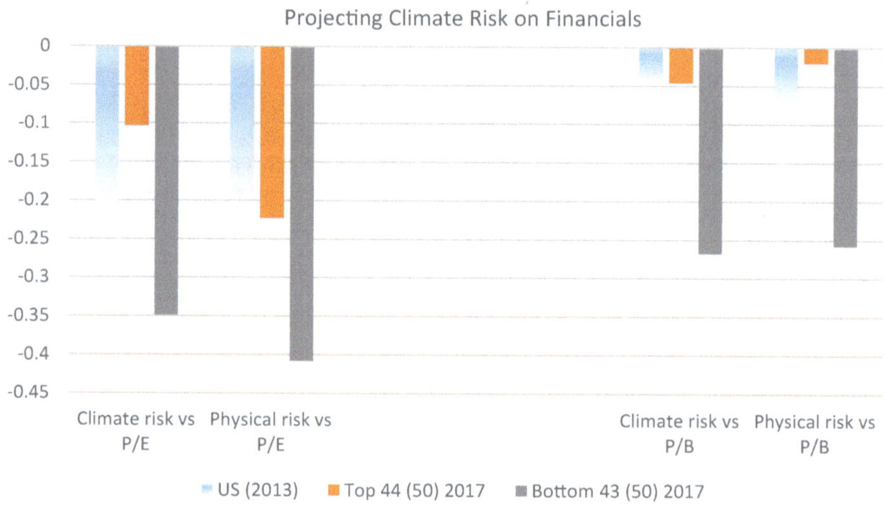

Fig. 4 Climate risk correlation with firm profitability ratios

[8]CDP 2018, "…over the past five years more than twice as many climate-related resolutions have been filed at oil and gas companies than in the preceding five years, and votes for shareholder resolutions relating to 2 °C scenario analysis more than doubled between 2014 and 2018".

The results are even more concerning when comparing top versus bottom companies. The correlation between climate risk and financial ratios is three times stronger for bottom 50 companies (from −0.1 to −0.35 for P/E ratio), highlighting the climate risk for companies that have low profitability and fail to disclose any climate risk data due to either lack of awareness of climate risk or dismissal as not relevant. It is therefore in the company's advantage to marginally increase the data transparency around climate risk and to minimise at least the risk of not understanding the climate risks. The comparison is even stronger when using the P/B ratio.

Interestingly, the correlation of climate risk to P/E and P/B ratios is equal or stronger for US companies that Asian companies are seating at the top 50 MSCI EM ETF. One possible explanation is a much sticker and more punishing regulatory environment in the USA than in Asia.

5 Climate Risk Awareness and Disclosure in Asia

Having established the relationship of climate risk to firm's financial profitability, the importance of data disclosure on climate risk and the dispersion of this disclosure for emerging market companies, we turn our focus on an intra-region Asian analysis, using a comprehensive unique set of 350 Asian companies that reported to CDP in 2017.

According to the CDP Hong Kong and South East Asian Report[9] 2017, 350 companies from Hong Kong, China, Indonesia, Thailand, Malaysia, Singapore and the Philippines are requested by CDP to fill in the questionnaires and scored accordingly. CDP nominal grade distribution (numeric CDP score not disclosed on the website) and the completeness of environmental disclosure on Bloomberg in each of the seven countries are manifested in the following bar charts (Fig. 5).

As expected among the 350 companies, 326 have been disclosing some information to get a CDP rating. However, 80% of them failing to provide sufficient information for evaluation hence get an F rating. Only HK, Thailand and Singapore have B-awareness (second highest), while none makes it to the A-leadership position (first highest). Concerning country performance, Singapore ranks the best where only 66% of companies fail to disclose information for evaluation. Indonesia and HK rank the lowest with almost 98% and 80% of companies failing to pass the minimum stage F-failure to disclose.

While CDP rating is used to calculate the transition risk, for the physical risk, we have to rely on complete environmental disclosure reported in Bloomberg as displayed in Fig. 6; hence to avoid misleading result due to incomplete reporting, our working sample size drastically shrinks from 326 to 85 companies only (see Table 2). This reduced sample set is the base for calculating the correlation between climate risk (as well as physical risk) and financial KPIs in the Asia region.

---

[9]Hong Kong and Southeast Asian Report 2017, CDP.

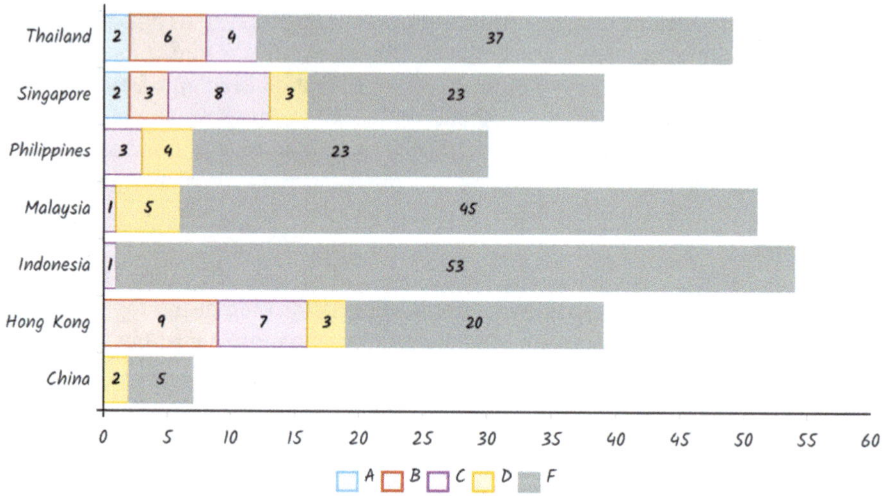

Fig. 5 CDP grade distribution for the 350 Asian companies reporting to CDP (A-leadership; B-management; C-awareness; D-disclosure; F-failure to provide sufficient information for evaluation)

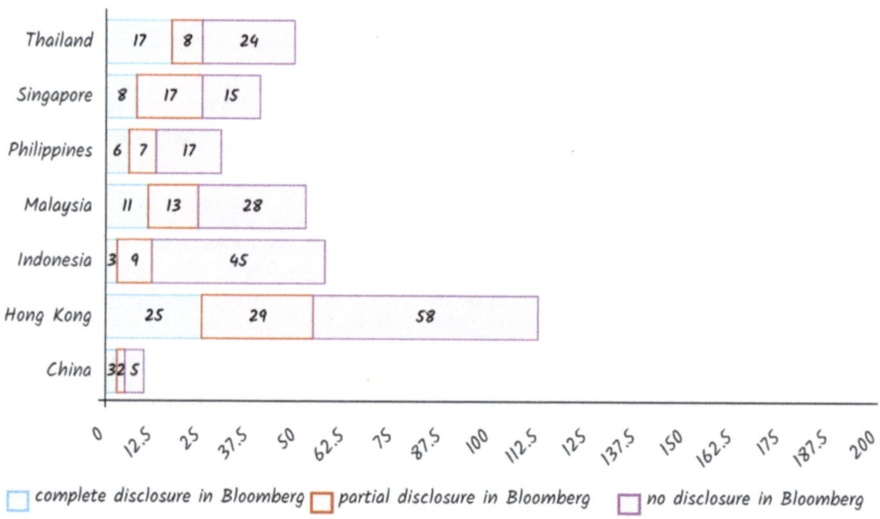

Fig. 6 Bloomberg disclosure level for the 350 Asian companies that report to CDP in 2017

General information of companies (market value and sector distribution) in different regions is briefly summarised in Tables 1, 2.

Hong Kong, Thailand and Malaysia have the largest share of companies reporting to both CDP and Bloomberg. Interestingly, the sectors that dominate across Asia are financials, industrials and consumer discretionary. Although not presented here, but

Table 1 Asia market coverage. In brackets is the % of companies for each sector relative to country sample

Region	Hong Kong	Malaysia	Singapore	Thailand	Philippines	China
Average market value	91,996 M HKD	25,141.16 M MYR	14,233.82 M SGD	304,762.23 M THB	312,549.20 M PHP	77,032.94 M RMB
Consumer discretionary	**24 (20.87%)**	**5 (9.80%)**	**3 (7.69%)**	**5 (10.20%)**	**1 (3.33%)**	**6 (46.15%)**
Consumer staple	1 (0.87%)	6 (11.76%)	9 (23.08%)	5 (10.20%)	4 (13.33%)	0 (0%)
Energy	2 (1.74%)	3 (5.88%)	1 (2.56%)	8 (16.33%)	2 (6.67%)	0 (0%)
Financials	**36 (31.30%)**	**12 (23.53%)**	**12 (30.77%)**	**9 (18.37%)**	**11 (36.67%)**	**0 (0%)**
Health care	1 (0.87%)	2 (3.92%)	0 (0%)	2 (4.08%)	0 (0%)	0 (0%)
Industrials	**15 (13.04%)**	**10 (19.61%)**	**9 (23.08%)**	**5 (10.20%)**	**7 (23.33%)**	**5 (38.46%)**
Information technology	5 (4.35%)	0 (0%)	1 (2.56%)	2 (4.08%)	0 (0%)	2 (15.38%)
Materials	7 (6.09%)	4 (7.84%)	2 (5.13%)	5 (10.20%)	0 (0%)	0 (0%)
Telecommunication service	3 (2.61%)	4 (7.84%)	2 (5.13%)	4 (8.16%)	2 (6.67%)	0 (0%)
Utilities	10 (8.70%)	5 (9.80%)	0 (0%)	4 (8.16%)	3 (10%)	0 (0%)
Companies with disclosure in CDP and Bloomberg	**25**	**11**	**8**	**17**	**6**	**8**

Table 2 A heat map for climate risk in Asia. It is the average physical, transition and climate risk per country given the number of companies that have full disclosure in CDP and Bloomberg. The darker the colour, the higher the risk is

Region	Average physical risk per company (based on Bloomberg data)	Average transition risk per company (based on CDP score)	Average total climate risk per company (Combined)
Hong Kong	6.48	2.42	8.90
Malaysia	4.91	2.72	7.63
Singapore	6.25	2.04	8.29
Thailand	7.82	2.32	10.15
Philippines	6.00	2.48	8.48
China	6.88	2.61	9.48

provided in Appendix, Table 4, consumer staples/discretionary and financials are the two sectors that report the most in climate-related issues.

Having collected the data on the company level, we attempt to represent the climate risk on a country level. Average physical risk, average transition risk and average total climate risk per company based in different regions are summarised in Table 3. As indicated in the table, the colour gradient displays the seriousness of different risks. The darker the colour, the higher the risk is. As shown, Thai companies have the highest physical risk, while Malaysian companies have the highest transition risk. While these results are by far not conclusive given the limited data set, they are partially indicative of the level of risk related to climate risk on company level given a specific country. Note that physical risk is measured only on the company level and not country-specific physical risks such as extreme weather and water scarcity. We do not comment on China as the sample size is extremely limited.

6 Climate Risks and Company Performance Relationship Beyond Regions

This section further explores the relation of climate risk to company valuation combining inter-region with an intra-region analysis. For better understanding and easier comparison, we include the numbers for MSCI EM ETF and US companies as shown in Fig. 4.

Our comprehensive analysis displayed in Table 3 leads us to conclude that: first, the complete disclosure rate on Bloomberg is much higher in the USA than in the Asian market, almost double. Second, CDP scoring A-range (leadership level) is much higher in the USA than in the Asian region. Many Asian companies still

Table 3 Climate risk correlation to valuation across regions and countries

Regions/countries	US	MSCI EMETF Top50	MSCI EMETF Bottom 50	China	Hong Kong	The Philippines	Thailand	Singapore	Malaysia
Climate risk awareness and disclosure									
Complete Bloomberg disclosure in %	49.26%	36.64%	18.6%	20%	22.32%	20%	34.69%	20%	21.15%
CDP scoring A-range in %	22.79%	29.55%	6.98%	0%	0%	0%	4.08%	5%	0%
Climate risk versus P/E and P/B ratios									
Correlation between climate risk and P/E	−0.2033	−0.1036	−0.3487	−0.2698	0.0886	0.805	−0.3818	0.7774	0.2705
Correlation between climate risk and P/B	−0.0449	0.0463	−0.2668	−0.5082	−0.0881	0.2179	−0.2978	−0.6437	−0.1881
Physical Risk versus P/E and P/B ratios									
Correlation between physical risk and P/E	−0.1981	−0.2226	−0.4081	−0.3055	0.0529	0.8097	−0.3174	0.7334	0.2619
Correlation between physical risk and P/B	−0.0694	0.0205	−0.257	−0.5481	−0.1485	0.3784	−0.2543	−0.731	−0.1995

receive an F-score from CDP, meaning failure to provide sufficient information for rating. Third, climate risk is overall negatively correlated with P/E and P/B ratios, indicating potentially hidden and unpriced risks. Fourth, this relationship is even stronger for bottom 50 MSCI EM ETF companies. Fifth, Thailand and China both display a consistent negative relationship between climate risk and firm valuations. Sixth, while the sign of the correlation and the magnitude are essential, their statistical significance is equally relevant for us to conclude the relationship. This is extremely important, especially for China, Singapore and the Philippines with only a handful of companies satisfying and disclosing climate risk as represented in Table 2.

7 Conclusions

As climate risk becomes more evident, more immediate and more accessible to translate into numbers, while governments introduce more stringent regulations to cope with the transition from high- to low-carbon economies, business-as-usual is no longer viable for companies. Our analysis shows that only a handful of companies among the already top-performing ones, measure, disclose or are aware of climate risks. Interestingly even among the top 50 companies part of MSCI EM ETF Index, there are strong leaders as well as laggards on climate disclosure, indicating that climate related risk is still far from being understood.

It is common practice to define climate risk as the sum of transition risk plus physical risk. Transition risk is related to the company level of awareness and management of climate risk. We calculate transition risk using CDP score as our input, meaning the higher the score, the lower the transition risk. CDP score is based on a comprehensive CDP questionnaire assessing a company's policies and targets an understanding of climate risk. Physical risk is a positive function of GHG emission, waste management and water consumption. In order to calculate all the above, we need CDP scores as well as Bloomberg data.

Among 350 Asian companies reporting to CDP, 326 have been disclosing some information to get a CDP rating. Nonetheless, 80% of them failing to provide sufficient information for evaluation hence get F-level rating. Our sample shrinks further as we require full disclose in Bloomberg as well.

Based on our simple analysis, our findings show that companies in Singapore, HK and Thailand seem to have the lowest transition risk while Malaysian companies rank as highest. The opposite seems true for physical risk, with HK and Thailand (and China) having the highest level, hence the highest probability that more stringent regulations or global warming-related risks such as drought or flooding will affect their operations and therefore their cash flows.

Lastly, we observe an overall negative relationship between climate risk (and physical risk) and company valuations captured by P/E and P/B ratios. This is an essential result for asset managers, and asset owns for valuations and portfolio allocations. Climate risk will no longer be a second-order risk. Once a hidden and extreme risk, climate risk, it should be considered a systematic risk and therefore demand a risk premium. Next step is to price climate risk [15].

Our analysis has obvious limitations arising from data shortage. More comprehensive data and non-linear regression analysis are needed to capture a more robust relationship between climate risk and firm performance. For practical use, this study can be easily extended to find the most resource-efficient companies in Asia—companies that generate more sales with less carbon, water and waste [5].

Acknowledgements We acknowledge our research assistant HUI, Yiting of HKUST, 2018, for handling the data and doing the analysis for this research project.

Appendix 1

CDP provides a scoring decision chart (see Fig. 1), allowing us to make this translation possible by referring to this scoring criteria. Firstly, we divided the full score (100) into five successive hierarchical groups. F-range is equivalent to a numeric range of 0–20, D-range is equivalent to 20–40, C-range is equivalent to 40–60, B-range is equivalent to 60-80, and A-range is equivalent to 80–100. According to CDP explanation, a company must reach the threshold of the fundamental level to be evaluated on the next level. For example, a company must score more than 80% regarding "disclosure" in order to be evaluated regarding "awareness". If the company is given a nominal grade of C, then its numeric score is estimated to be approximately $40 + (60–40) * (0 + 44\%)/2 = 44.4$. For another example, if the company is given a nominal grade of A, then its numeric score is estimated to be approximately $80 + (100 - 80) * (80\% + 100\%)/2 = 98$ [6].

Appendix 2

See Table 4.

Table 4 Asian companies that report to CDP and have Bloomberg data on physical risk as of June 2018

Company	Country	Business sector	Climate disclosure CDP score
Li Ning Company Ltd	China	Consumer discretionary	F
Alibaba Pictures Group	Hong Kong	Consumer discretionary	F
Brilliance China Automotive Holdings Ltd	Hong Kong	Consumer discretionary	F
Melco Crown Entertainment Ltd	Hong Kong	Consumer discretionary	F
Melco International Development Ltd	Hong Kong	Consumer discretionary	F
Nan Hai	Hong Kong	Consumer discretionary	F
Samsonite International SA	Hong Kong	Consumer discretionary	F
Sands China Ltd	Hong Kong	Consumer discretionary	F
Techtronic Industries	Hong Kong	Consumer discretionary	F
Yue Yuen Industrial	Hong Kong	Consumer discretionary	F
Global Brands Group	Hong Kong	Consumer discretionary	C
AIA Group Ltd	Hong Kong	Financials	C
Bank of East Asia Limited	Hong Kong	Financials	D
First Pacific	Hong Kong	Financials	B
Swire Properties	Hong Kong	Financials	B
BOC Hong Kong	Hong Kong	Financials	F
China Jinmao Holdings Group Limited	Hong Kong	Financials	F
Hongkong Land Holdings	Hong Kong	Financials	F
Hong Kong Aircraft Engineering	Hong Kong	Industrials	B
Jardine Strategic	Hong Kong	Industrials	F
Semiconductor Manufacturing International Corp	Hong Kong	Information technology	C
Fosun International	Hong Kong	Materials	F
Xiwang Special Steel Co Ltd	Hong Kong	Materials	F
Hong Kong and China Gas Company Limited	Hong Kong	Utilities	F
China Power International Development Limited	Hong Kong	Utilities	F
Power Assets Holdings Limited	Hong Kong	Utilities	F

(continued)

Table 4 (continued)

Company	Country	Business sector	Climate disclosure CDP score
Bank Danamon Indonesia Tbk	Indonesia	Financials	F
Kalbe Farma Tbk PT	Indonesia	Health care	F
Perusahaan Gas Limited	Indonesia	Utilities	F
Astro Malaysia Holdings	Malaysia	Consumer discretionary	F
Berjaya Sports Toto Berhad	Malaysia	Consumer discretionary	F
Genting Berhad	Malaysia	Consumer discretionary	F
IOI	Malaysia	Consumer staples	F
PPB Group	Malaysia	Consumer staples	F
Sapura Kencana Petroleum Bhd	Malaysia	Energy	F
AirAsia Berhad	Malaysia	Industrials	F
Westports Holdings Berhad	Malaysia	Industrials	F
Lafarge Malayan Cement Bhd	Malaysia	Materials	F
Samling Global	Malaysia	Materials	F
Tenaga Nasional	Malaysia	Utilities	D-
Universal Robina	Philippines	Consumer staples	D
LT Group Inc	Philippines	Consumer staples	F
Puregold Price Club, Inc	Philippines	Consumer staples	F
Megaworld Corp	Philippines	Financials	F
Metropolitan Bank & Trust	Philippines	Financials	F
Energy Development Corp	Philippines	Utilities	C
Mewah International Inc	Singapore	Consumer staples	F
Permata Hijau Group	Singapore	Consumer staples	F
BW LPG	Singapore	Energy	F
CapitaLand Commercial Trust	Singapore	Financials	B
City Developments Limited	Singapore	Financials	A-
DBS Group Holdings	Singapore	Financials	C
Keppel Land Company Limited	Singapore	Financials	C
UOL Group	Singapore	Financials	F
Sembcorp Marine	Singapore	Industrials	F
Broadcom Limited	Singapore	Information technology	F

(continued)

Table 4 (continued)

Company	Country	Business sector	Climate disclosure CDP score
BEC World Public Company Ltd	Thailand	Consumer discretionary	F
Home Product Center, Plc	Thailand	Consumer discretionary	F
Robinson Department Store	Thailand	Consumer discretionary	F
Minor International PCL	Thailand	Consumer discretionary	C
Charoen Pokphand Foods PCL	Thailand	Consumer staples	B
CP ALL PCL	Thailand	Consumer staples	F
Thai Beverage PCL	Thailand	Consumer staples	F
Thai Union Group PLC	Thailand	Consumer staples	F
Central Pattana Pub Co Ltd	Thailand	Financials	F
Krung Thai Bank Pub Co Ltd	Thailand	Financials	F
Bumrungrad Hospital PCL	Thailand	Health care	F
Delta Electronics (Thailand) Plc	Thailand	Information technology	B
PTT Global Chemical	Thailand	Materials	A-
True Corporation	Thailand	Telecommunication services	C
Intouch Group	Thailand	Telecommunication services	F
Electricity Generating Public Co Ltd	Thailand	Utilities	F
Ratchaburi Electricity Generating Holding Public Company Ltd	Thailand	Utilities	F

References

1. Andersson M, Bolton P, Samana F (2016) Hedging climate risk. Financ Anal J 72(3):13–32
2. BP Statistical Review of World Energy Report (2018) https://www.bp.com/content/dam/bp/en/corporate/pdf/energy-economics/statistical-review/bp-stats-review-2018-full-report.pdf. Accessed Jan 2019
3. Basu S (1977) Investment performance of common stocks in relation to their price-earnings ratios: a test of the efficient market hypothesis. J Finance 32(3):663–682
4. Basu S (1982) "The relationship between earnings" yield, market value and return for NYSE common stocks. J Financ Econ 12(1983):129–156
5. BlackRock (2016) Adapting portfolios to climate change—implications and strategies for all investors. Blackrock investment institute
6. CDP (2017) Hong Kong & South East Asia Report. CDP
7. CICERO (n.d.) Center for international climate research. Retrieved from categorising climate risks for investors: https://www.cicero.oslo.no/en/CF-transitional-risk
8. Caldecott B, McDaniels J (2014) Stranded generation assets: implications for European capacity mechanisms, energy markets and climate policy Working Paper
9. Ceres (2018) SEC Sustainability Disclosure Sear Tool. https://tools.ceres.org/resources/tools/sec-sustainability-disclosure/@@ceres-search-s3
10. Deutsche Bank (2017) Measuring Physical Climate Risk for Equity Portfolios
11. Hong H, Li FW, Xu J (2017) Climate risks and market efficiency. Forthcoming J Econometrics
12. Howard A, Hassler M (2018) How will the physical risks of climate change affect companies? Schroders. https://www.schroders.com/en/insights/economics/how-will-physical-risks-of-climate-change-affect-companies/. Accessed Jan 2019
13. Jong M, Nguyen A (2016) Weathered for climate risk: a bond investment proposition. Financ Anal J 72. https://www.cfapubs.org/author/de+Jong%2C+Marielle, https://www.cfapubs.org/loi/faj
14. KPMG (2017) The KPMG survey of corporate responsibility reporting. https://home.kpmg.com/xx/en/home/media/press-releases/2017/10/companies-not-noting-financial-risk-of-climate-change.html
15. Litterman R (2018) Pricing climate change risk appropriately. Financ Anal J 67(5):4–10
16. McKinsey (2018) How climate change could affect corporate valuations. https://www.mckinsey.com/business-functions/strategy-and-corporate-finance/our-insights/how-climate-change-could-affect-corporate-valuations
17. Schroders (2018) Climate change: the forgotten physical risks. https://www.schroders.com/en/sysglobalassets/digital/insights/2018/thought-leadership/climate-change---the-forgotten-physical-risks_final.pdf/
18. Stephen BJ (2016) Climate Risk. Financ Anal J 72:3, 9–10. https://doi.org/10.2469/faj.v72.n3.5
19. Task Force on Climate-related Disclosures. https://www.fsb-tcfd.org/
20. Trinks A, Ibikunle G, Mulder M, Scholtens B (14 Sept 2017) Carbon intensity and the cost of equity capital. Available at SSRN: https://ssrn.com/abstract=3035864 or https://doi.org/10.2139/ssrn.3035864

Challenges and Countermeasures in the Development of China's Green Bond Market

Jingyan Fu, Shuyu Li and Artie W. Ng

Abstract After 30 years of economic reform and liberalization, China's economic and social development have achieved noticeable success whereas the consideration of sustainable development has become salient. While China has become an engine for the development of the global economy, its domestic economic development has advanced with noticeable environmental impairments. China now needs to strike a balance between economic development and environmental protection. Public funds alone are not sufficient for solving China's environmental and climate problems. The People's Bank of China has clearly stated that public finance can only meet 15% of the funding needed. Green bonds are becoming a powerful tool for mobilizing the global bond market to meet the demand for green investment. At present, green bonds account for less than 0.2% of the global bond market (2% in China) but have enormous potentials for expansion. This paper articulates the current development status of China's green debt market, the development experience of the international green debt market, issues associated with the development of China's emerging green debt market, and the corresponding countermeasures. This study conducts an overview of China's policy for the green debt market as well as its developmental challenges and opportunities.

Keywords Green bonds · Third-party certification · National strategy · Regional policy

1 Introduction

Studying China's green bond market requires some background knowledge of the emergence of green financing as an international initiative. In 2007, the European

J. Fu (✉) · S. Li
Institute of Resource, Environment and Sustainable Development, College of Economics, Jinan University, Guangzhou, China
e-mail: tfjyan@jnu.edu.cn

A. W. Ng
School of Professional Education and Executive Development, College of Professional & Continuing Education, The Hong Kong Polytechnic University, Kowloon, Hong Kong SAR, China

J. Fu and A. W. Ng (eds.), *Sustainable Energy and Green Finance for a Low-carbon Economy*, https://doi.org/10.1007/978-3-030-35411-4_3

Investment Bank took the lead in issuing climate-aware bonds. This was the first environmentally themed bond issued by a multilateral institution and set a precedent for green bond issuance. That is, clear funds would be used for green projects, and strict special funds would be set. Shortly thereafter, in 2008, the World Bank issued the world's first green bond, raising funds specifically for projects focused on climate change mitigation and adaptation. Since then, multilateral institutions, governments, and firms have been increasingly involved in the issuance of green bonds.

As the types and scale of green bond issuance have continued to expand, various organizations have defined it differently. The World Bank has defined green bonds as a fixed-income ordinary bond that provides investors with the opportunity to finance climate change mitigation and adaption by investing in green projects. The Organization for Economic Co-operation and Development (OECD) has defined green bonds as fixed-income securities issued by governments, multinational banks or firms to raise the necessary funds for projects that promote a low-carbon economy and adaption to climate change.

Climate Bonds Initiative, the most authoritative non-profit organization in global climate bonds and green bonds, has considered green bonds to be fixed-income financial instruments for raising funds for sustainable development or environmental projects. The Green Bond Principles have provided the most authoritative definition: the green bond is a bond instrument with benefits that should be applied to funding or refinancing new or existing green projects exclusively, by prior clear revenue use, direct project exposure or bond securitization. Green projects are defined as projects or activities aimed at addressing climate or other environmental sustainability issues. Although the definitions of the green bonds have differences, the core meaning of green bonds is largely similar as a bond-type instrument with fixed income applied to green projects.

This chapter provides a brief literature review on the development of green finance and green bonds from an international perspective. Special issues pertinent to the development of green bond market in China over the recent years are examined. Limitations, challenges, and opportunities for its sophistication are then articulated.

2 Literature Review

2.1 The Introduction and Development of Green Finance

Environmental issues have become a major problem in the sustainable development of humanity around the world. Different fields have been devoted to improving the environment through various ways. The financial sector, as the main driving force of economic development, also has attempted to contribute in its own way. Environmental finance is a consideration of environmental impacts in the financial decision-making process. Cowan [1] defined the concept of environmental finance and analysed its possible development pathways. Jeucken [2] studied the development of sustainable finance in the banking industry and asserted that the banking industry has minimal direct environmental impact.

Thompson and Cowton [3] asserted that a bank's loan business is affected by the natural environment and society's attention to environmental issues. Public policies and consumer environmental information need to be fed back into company operations, which in turn are related to bank credit. He empirically studied the relationship between bank credit and demand for environmental information. Ng [4] stated that the continued development of China's green financial system and complementary, emerging global financial centre demonstrates an encouraging international trend of greater investments into green and sustainable energy infrastructure projects.

2.2 *The Rise of Green Bonds*

In recent years, scholars have become widely concerned with in-depth comparisons of green bonds. Wood and Grace [5] argued that the growth in green bond sectors would depend on a variety of factors, including the ability of the issuers to undertake green projects, the size of the issue, the stakeholder involvement in setting the metrics, and the investor's funds on the bond.

Kaminker and Stewart [6] suggested that most economies of OECD countries had been experiencing low interest rates in the context of stable GDP growth. Therefore, rational investors had been more willing to invest in non-traditional financial products to obtain relatively safe and reliable income. The support policies for China's green bond market predate those in other countries. The policies that have entered the official approval stage include the exclusion of green loans from the calculation of loan-to-deposit ratios; the application of 75% of the preferential risk weight and capital supervision requirements, allowing financial institutions to provide pre-tax provisions for green credit assets; and preferential capital risk ratios for banks investing in green bonds, with 50% downward adjustments on their green bond risk assets.

Some scholars have studied the environmental and social performance of Green China Bank. There also have been many scholars who believe that green credit policies and other initiatives integrating the financial sector into social and environmental sustainability financing [7] may have a positive impact on sustainable development [8]. In addition, green credit could encourage banks to actively participate in the integration of environmental risks into their credit risk assessment procedures [9], so promoting cultural policy changes would move Chinese banks into a more sustainable direction [10].

3 Development Status of China's Green Bond Market

The green bonds issued by Chinese firms grew rapidly in 2016, from almost zero to 238 billion yuan (about US$36.2 billion), accounting for 39% of the global issuance [11]. The Agricultural Bank of China in London issued the first green bonds in China at the end of 2015. In December 2015, the People's Bank of China published the announcement of green financial bonds in the inter-bank bond market, China's current largest bond market. A few weeks later, the China Development and Reform Commission also announced guidelines for green corporate bond issuance. Since then, China's green bond market has grown rapidly, mainly due to the green bonds from large issuers such as Shanghai Pudong Development Bank, Industrial Bank, and Bank of Communications. As the sizes of the issues have grown, so have the variety of green bonds, which have included green asset-backed bonds (ABS) and Bank of China's green guaranteed bonds.

By the end of November 2018, more than 500 billion US dollars of global green bonds had been issued, and China had become one of the largest distribution sources in the global green bond market. Industrial Bank has played an important role in the domestic green bond market. Industrial Bank landed its first green financial bond in 2016. In November 2018, it successfully completed the issuance of green financial bonds in both domestic and overseas markets. The stock of green financial bonds exceeded 110 billion yuan, making it the commercial financial institution with the largest balance of green financial bonds issued at that time.

3.1 Types of Bonds in China's Green Bond Market

China has issued green bonds since 2015. By sorting out the green bond types in China published in the Wind database, we find that in China's green debt market, asset-backed bonds occupy a dominant position, accounting for 70%, while the shares of corporate bonds and financial bonds are relatively similar, at 14% and 16%, respectively (see Fig. 1).

3.2 The Issuers of China's Green Bond Market

The issuers in the Chinese bond market are mainly state-owned enterprises (Fig. 2). Among these, central state-owned enterprises have the highest amount of bond issuance, accounting for more than half of the total, while local state-owned enterprises account for 30% (see Fig. 2). Although the issuers in China's green debt market have become increasingly diversified, state-owned enterprises still occupy an absolute leading position, while private enterprises and wholly owned enterprises still have room for potential development space.

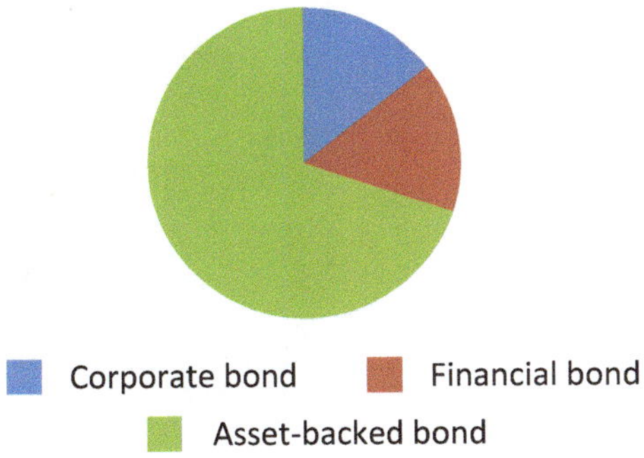

Fig. 1 Types of bonds issued within China's green debt market. *Data source* Wind (Wind is a leading financial data, information, and software services company in mainland China. In the Chinese market, Wind's clients include most of China's securities companies, fund management companies, insurance companies, banks, and investment companies. In the international market, qualified foreign institutional investors (QFII) have been approved by the China Securities Regulatory Commission. Many of these qualified institutions are Wind's customers. A large number of Chinese and English media, research reports, and academic papers often refer to Wind Data.) database. https://www.wind.com.cn/NewSite/edb.html

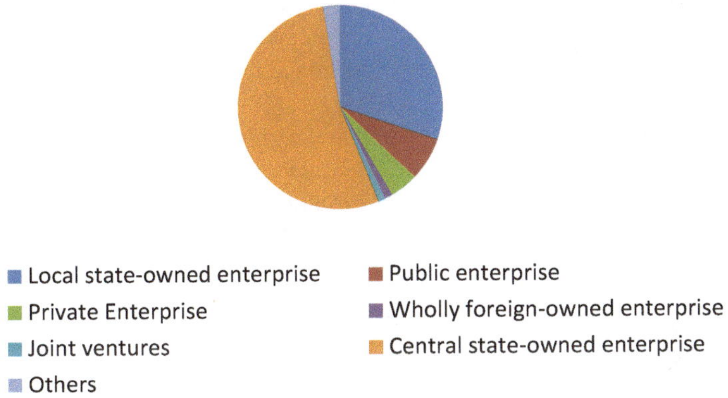

Fig. 2 Issuers in China's green bond market. *Data source* Wind database. https://www.wind.com.cn/NewSite/edb.html

3.3 *China Green Bond Market Rating*

Green bonds have a wide range of subject ratings, with a minimum rating of A+ and a maximum of AAA. Among them, the AAA rating is the most common, accounting for 65%, and the rating of A+ comprises only 0.3% (see Fig. 3). This indicates that most of the bonds issued by the China's green bond market have received high standard ratings. Through the participation or endorsement of high-confidence third parties, product credibility can be increased, which will help reduce moral hazard and ensure the interests of investors.

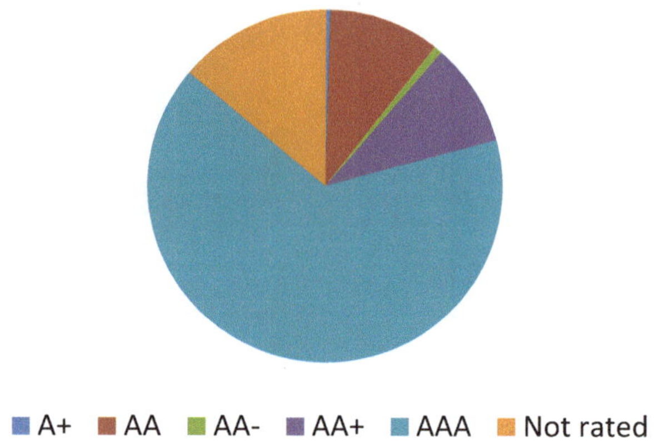

Fig. 3 China's green debt market bond rating. *Data source* Wind database. https://www.wind.com.cn/NewSite/edb.html

3.4 *Main Distribution Areas of China's Green Bond Market*

Geographically, the distribution of China's green bonds has been uneven across China. Figure 4 shows the regional distribution since the first issuance of green debt in China. This highlight focuses on Beijing, Shanghai, Guangdong, Jiangsu, and Zhejiang, which markets have relatively developed in China. Through the analysis of the data, we find that green debt is most concentrated in the East China region, represented by Shanghai.

Moreover, other relevant data indicate that the green debt issuance in Beijing and Guangdong Province in the first half of 2018 demonstrated favourable momentum in issuing green bonds (Fig. 5). Guangdong Province is one of the five green financial pilot zones approved by the State Council. Since 2017, its Huadu District has taken

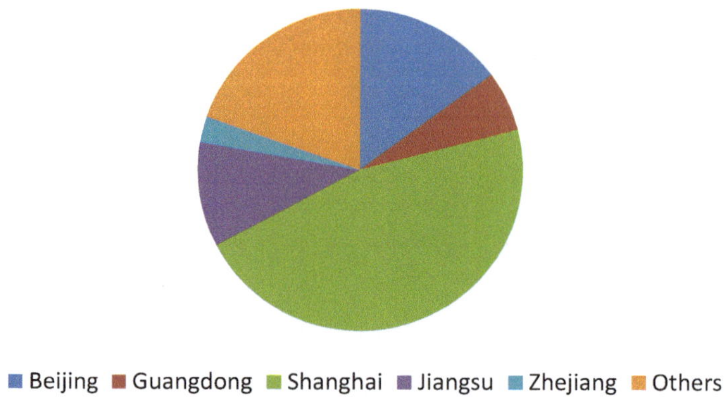

Fig. 4 Major distribution areas of China's green bond market. *Data source* Wind database. https://www.wind.com.cn/NewSite/edb.html

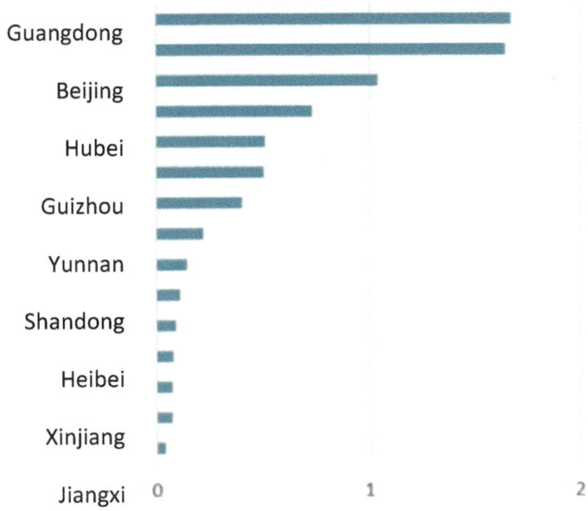

Fig. 5 Green bond issuance in the third quarter of 2018 by province (US$1 billion). *Source* China Green Bond Market 2018 Quarterly Report. https://www.climatebonds.net/newsletter/china

the lead in launching the "1 + 4" green financial policy. The region will arrange special funds of at least 1 billion yuan for five consecutive years and support the development of green finance through tools such as green bonds (Fig. 5).

3.5 Investment Direction of Raised Funds in China's Green Bond Market

According to Climate Bond Initiative's definition, renewable energy has the greatest potential for fundraising. As non-financial companies, CGN International and China Energy Conservation Wind Power Co. Ltd. both have invested in green bond to fund wind farm projects. Since the beginning of China's green bond market in early 2016, low-carbon transportation has been ranked in the top three areas for fundraising. Qualified assets have included subway and bus infrastructure. For example, the Urumqi City Transportation Group has issued 1.5 billion yuan (US$220 million) in green bonds to be used for bus lane construction, bus stops, bus signal systems, BRT, and bus parking facilities. Beijing Infrastructure Investment Group and Wuhan Metro also have used green bonds to support local subway construction.

Since the beginning of 2018, the proportion of green bond funds in the water sector has continued to increase, from 7% in the first quarter to 18% in the third quarter. The uses for raised funds in the water sector have been diverse, from wastewater treatment to rainwater harvesting and distribution. As the number of extreme rainfall events has increased, we have observed increasing frequency of urban flooding. Therefore, we should encourage the rational design and management of stormy water drainage systems and infrastructure to ensure the climate adaptability of urban infrastructure (Fig. 6).

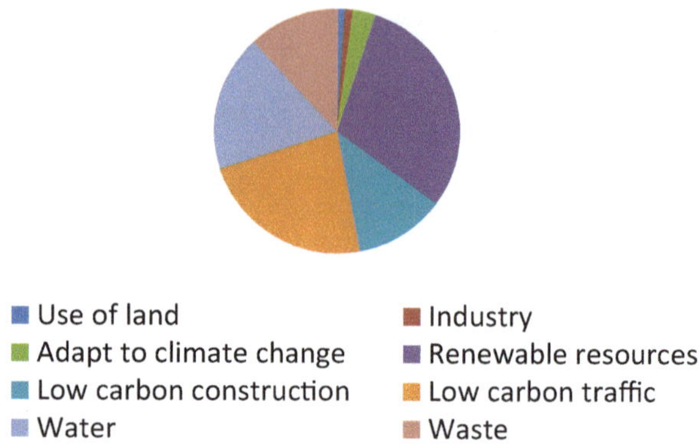

Fig. 6 Use of funds raised in the green debt market in the third quarter of 2018. *Source* China Green Bond Market 2018 Quarterly Report. https://www.climatebonds.net/newsletter/china

4 Development Features of Overseas Green Bond Markets

The development of the green bond market began in 2007–2008 with the start of green bond issuance by the World Bank and the European Investment Bank. From 2007 to 2012, the green bond market was dominated by development banks such as the European Investment Bank, the International Finance Corporation, and the World Bank. In 2013, corporate entities issued their first green bonds, which spurred more private-sector participation, including corporate and commercial banks. As a set of voluntary guidelines, the Green Bond Principles (GBP) were released in early 2014, driving market best practices in transparency and reporting.

The issuance of the first corporate green bond and the Green Bond Principle has had a strong catalytic effect on the further development of the market. The market size of green bonds has increased from US$3 billion in 2012 (approximately 21 billion yuan) to US$81 billion in 2016 (approximately 559 billion yuan), covering 14 of the G20 countries. As the size of the market has grown, the issuers and investors of green bonds also have demonstrated diverse trends. In 2016, global green bond issuance reached US$81 billion (approximately 559 billion yuan), which was largely driven by Chinese issuers. In general, the characteristics of the foreign green debt market can be summarized as follows.

4.1 *The Development of Green Bond Standards*

The rise of green bonds in the international market has been the inevitable result of economic development. The concept of international investor responsibility has become increasingly popular, as people have been paying greater attention to climate change and environmental issues. This change in investor value judgement has become an important driving force for the development of green debt. Based on the existing market, issuers, investment institutions, and underwriters have jointly endorsed the guidelines for voluntary participation.

To enhance the sustainability and transparency of green bond information disclosure and promote the healthy development of the green bond market, the GBP Executive Committee and International Capital Market Association (ICMA) jointly launched the GBP in 2014. Based on these voluntary guiding principles, by 2015, there were more than 100 members, five including green bond issuers, underwriters, and investors. Under the GBP, CBI has developed a complementary standard which aims to provide assurance that the usage of raised funds is consistent with low-carbon economy requirements. The rapid development of the foreign green debt market has been related to the voluntary nature of these principles. The voluntary equality of all parties has fully mobilized the subjective initiative of the participants. The multilateral efforts towards one goal have greatly reduced communication costs and time costs (Wang 2018).

4.2 *Government Policy Support*

To promote the development of green bonds, many countries have begun to formulate various preferential policies, including green loan discounts and price subsidies. Currently, more than 60 countries have formulated price subsidy policies. The long-term repurchase of enterprises' products according to fixed prices has largely ensured the flow of funds of such enterprises and the attractiveness of the bonds.

4.3 *International Green Bond Certification System*

International mainstream certification systems include the Green Bond Principles (GBP) and the Climate Bonds Standard (CBS). The GBP recommend that issuers use external certification to ensure that bonds issued are consistent with green bonds. Definitions, requirements, and the types and levels of certification are encouraged to be included in professional assistance from consultants to review or assist in the establishment of project evaluation and selection systems. Independent audits by third parties should be based on the second-party standards. The issuance of the "second opinion" CBS requires that a third-party reviewer must be appointed and allowed to provide an assurance report on whether the bond meets the climate bond standard. Presently, the domestic standards for green bonds do not distinguish between second-party assessments and third-party certifications, but international standards for green bonds (such as GBP) provide a more detailed description of second-party assessments.

The certification of green bonds should not be limited to the classification of projects, but rather should further reveal the expected or actual environmental benefits of the projects they finance, in order to avoid the suspicion of so-called greenwashing. Presently, China has not yet formed a mature certification mechanism, which requires that third-party institutions establish a more professional and quantitative evaluation model, in order to better assess the green project implementation and effect of the enterprise.

4.4 Enhancing the Information Disclosure System

To promote the development of the green bond market, the exchange has developed many plans to help achieve sustainable development. In 2014, the World Federation of Exchanges established a special working group for the development of green bonds. According to its research, many large companies have participated in the above projects. The emergence of various indexes has provided very important references for strengthening the transparency of green assets and has effectively promoted its commoditization process while enhancing its expected benefits.

5 Challenges with China's Green Bond Market

Having developed earlier, the foreign green bond system is more comprehensive in terms of the definition of green projects, the review by external auditing agencies, and the evaluation of third-party certification bodies. For the development process of China's green bonds, given its special national conditions, lessons could be learned from advanced foreign systems to reduce exploration time, make the domestic green bond market healthier, and enable more green enterprises to use capital efficiently. At the same time, non-green enterprise's development of green projects could also be financed through green bonds, allowing social capital to contribute to the protection of the ecological environment. China is in a period of rapid development of green bonds, and there are still some shortcomings in the formulation of standards, the normative nature of the external audit industry, the development of potential investors, and the encouragement of government policies. To make up for these shortcomings, China's financial system needs to be strengthened, and more enterprises should be encouraged to fulfil their social responsibilities while maintaining their own steady development and growth, contributing to China's economic development and ecological protection.

5.1 Green Bond Policies

The development of the international green bond market has been mainly spontaneous. In contrast, the development of China's green bond market has been mainly guided by regulatory authorities. The issuance of green bonds in the Chinese domestic bond market requires the approval of relevant regulatory authorities, such as the People's Bank of China, the National Development and Reform Commission, and the China Securities Regulatory Commission. The approval of green bonds requires meeting guideline requirements, which include those for qualified green projects, fundraising, information disclosure, and external review. The green bonds of different regulators have varying priorities. In China, the most authoritative green debt guidance is the "Green Bond Support Project Catalogue" of the People's Bank of China and the "Green Bond Issuance Guidelines" of the National Development and Reform Commission. However, there are significant differences in the policy requirements for green bond issuance.

The People's Bank of China's guidance focuses on defining green project categories and has more detailed requirements such as project division, condition definition, fundraising management, information disclosure, and third-party certification. It is applicable to the inter-bank and exchange markets. The guidelines focus on clarifying the key areas of policy support for green projects, giving directional guidance, and proposing a number of specific measures in terms of incentives.

5.2 *Market Participation*

China's green bonds started relatively late, and its development is still in its infancy. On the whole, the issuers are still mostly single, and market participation is still not high. Specific performance characteristics are as follows:

First, the main body issuing green bonds in China has certain limitations. The green debt issuance standard is stricter, with higher standards for the issuer than those for ordinary bonds. Therefore, the market entry threshold is much higher than the ordinary creditor's right, which is unfavourable to expanding the market rapidly. Second, the enterprises' enthusiasm to choose green debt financing is not high. Enterprises' enthusiasm for green corporate bond selection is not high, mainly because there are two major differences between green corporate bonds and ordinary bonds. First, green corporate bonds must raise funds for green projects that meet regulatory standards. Second, they are earmarked. To ensure that funds are used for green projects, special accounts need to be set up to supervise funds. The funds raised can be used for project construction expenditures and can also be used to repay project loans and supplement project working funds. In terms of information disclosure and material review, the requirements for green corporate bonds are much higher than for ordinary bonds. They are also subject to supervision from multiple authorities. Furthermore, companies must cover the costs of completing the assessment and certification, which undoubtedly increases their financing and time costs. All of the above would reduce the enterprises' enthusiasm to choose green corporate bonds when issuing bonds.

Finally, market investors are not willing to invest in green bonds. Issuers, investors, and the market investment supporting environment are the main managers of investment products in the capital market. Green bonds are no exception. In China, the promotion of green bonds is mainly led by the government from the top down. In this development process, the emphasis has been mainly on the issuers' economies of scale, while lacking spontaneous green investor teams. Market investors have been more concerned about the issuer when choosing investments; whereas, the focus on green attributes has not been sufficiently prominent, and the awareness of green investment has been weak.

5.3 *Green Corporate Bond Issuance and Third-Party Certification*

Green bonds should be certified by third-party certification bodies or independent second opinion institutions. This is the difference between green bond issuance and other bond issuance, and it is a distinctive feature of green bond issuance. To date, most of the green bonds issued in China have involved third-party certification. China has its own local third-party certification body but lacks an independent green certification body. To improve the credibility of green corporate bonds, the introduction of third-party certification is an indispensable step to attract investors. When China's first green corporate bond, 16 Beijing Auto Green Bond 01, was issued in April 2016, the NDRC did not hire a third party after the internal environmental division, and the industry division evaluated the green compliance of the bond support project. These organizations conduct certification.

From April 2016 to August 2017, a total of 70 green bonds were issued in China, of which 49 were third-party certified green bonds, accounting for 70% of the total issued amount. Of the 21 green bonds that lacked third-party certification, 14 were green corporate bonds, accounting for only 66.67% of the uncertified bonds. The main reason for this phenomenon is that the corporate debt supervisor has not proposed a mandatory requirement for third-party certification in the regulatory policy. In the "Guidelines for the Issuance of Green Bonds" issued by the National Development and Reform Commission on 31 December 2015, there are no special requirements for the disclosure of information on green corporate bonds. The issuing companies only need to follow the information disclosure rules for corporate bond issuance. The guidelines for general corporate bond information disclosure can be used.

5.4 *Development of Relevant Laws and Regulations*

Based on international experience, the regions with rapid development of green bonds have matured into effective systems. In the process of developing green finance, the development of carbon finance has been crucial. China has also established relevant pilot programs, but there are still some problems in the development of green finance in the pilot cities in terms of legal systems and business rules. These problems have been associated with domestic financial institutions' unfamiliarity with the carbon finance trading model, which has been at a relatively low level of development for a long time, and the results achieved have been very limited.

The definition of a green bond is very basic, but difficult to concisely articulate because there presently is no unified, global definition of a green bond. Due to the lack of a unified definition, regulatory standards are not uniform. Fundamentally, the lack of effective regulation may lead to disorderly competition. The risk caused by inconsistent definitions is not considered the main risk. The major drawback is that the issuer of the green bond may provide the potential investor with misleading or insufficient information that could influence the final investment decision of the rational investor. Due to the lack of relevant mandatory regulations and standards, issuers appear to enjoy a higher degree of freedom in interpreting green bonds. Therefore, for investors, the burden of proof is heavier than that of investing in non-green bonds.

6 Concluding Notes

6.1 Importance of Defining Green Projects

Presently, there is a unified global consensus on the definition of a green bond. On 27 March 2015, the International Capital Market Association (ICMA) and 130 other financial institutions jointly described the principle of green bonds, stating that green bonds refer to any funds that are used exclusively to finance or meet the requirements of green projects. Green projects are those that promote environmental sustainability and are evaluated and selected by the issuers and relevant institutions. Green projects include the climate change mitigation and adaptation, curbing natural resource depletion, biodiversity conservation, and pollution control. Although China is yet to provide an explicitly comprehensive definition of green bond, relevant regulatory authorities are studying and formulating the "Green Bond Issuance Management Measures". It is expected such a global movement would promote the formation of a better definition for green bonds issued in China, mainly by considering which green items meet the prescribed conditions, and by determining whether existing green projects in China deviate from the international Green Bond Principles.

6.2 Limiting the Distributions of Green Bond Funds

Issuers can use green bond financing to fund new and existing green projects, invest in specific green projects, issue green loans, or simply invest in general green funds. All identified green project categories should provide clear environmental benefits that can be described, quantified, and assessed if feasible. In addition, some issuers have begun to specifically exclude specific business activities. For example, the African

Development Bank, the European Bank for Reconstruction and Development, and the European Investment Bank have banned green financing for nuclear or coal operations.

6.3 Strengthening the Management of Green Bonds

The net financing proceeds of green bonds are typically transferred to a secondary portfolio or tracked and managed by the issuer through other means as part of its loan and investment business. Once a qualifying green project is identified, funds should be immediately disbursed to the issuer's primary account for distribution to specific projects. However, if the use of funds is partially included in the final clause and is not differentiated from other debt instruments, there would be no guarantee that all funds would be used for green bond investments. If the money is distinguished from other debt instruments and managed separately—for example, as a secondary investment portfolio or a special fund, it can be tracked and managed within a special account.

6.4 Attaching Third-Party Certification to Green Bonds

In addition to their main credit rating and bond credit rating, green bond issuers must also make their green characteristics credible and persuasive for investors. The common practice globally is to request an independent professional certification body to issue a green certification for the direction of use for raised funds, a so-called second opinion. The second opinion has a detailed description of the investment in green bond proceeds, which can enhance the transparency of green bond information disclosure and attract more investors. Presently, internationally authoritative second opinion providers include International Centre for Climate and Environmental Research (CICERO), Vigeo Rating, DNVGL Group, Climate Bond Committee (CBI), Oekom Research Centre, KPMG, Sustainalytics, and Trucost Wait. As of the end of 2014, approximately 63% of the 300 green bonds issued globally had international green certification, and 54.22% of the issuers chose CICERO for second opinions. If China intends to issue green bonds, the second opinion certification also would be an indispensable step. To date, such independent green certification institutions have not been established in China.

6.5 Flexible Design of the Structure of Green Bonds

International green bonds are basically classified into four bond structure categories: (i) specific income use green bonds, (ii) specific income use green guarantee bonds,

(iii) green project bonds, and (iv) green asset support bonds. Investors with specific income use green bonds have full recourse to the issuer, so the green bond has the same credit rating as the other bonds issued by the issuer. The issuer uses the proceeds of the bond to support the sub-asset portfolio of the green project and stipulates the scope of use and setting up of internal mechanisms for tracking and reporting. Most of the green bonds issued by international financial organizations adopt this structure.

The investors of the specific income use green income guarantee bond have no debt recourse right to the issuer. The issuers use the income and tax credits obtained from the project operation as the bond guarantee. The issuers track and report on the use of bond proceeds. Most municipal bonds adopt this structure. Investors in green project bonds have debt recourse limited to specific project assets, i.e. investors are directly exposed to project risks. Investors of green asset support bonds can make debt recourse to one or more specific projects that are grouped together, including asset-backed bonds (ABS) and other structured products, which are generally backed by cash flows generated from operating assets.

6.6 Reliance on National Strategies and Regional Policies

6.6.1 National Strategies

To encourage investors to purchase green bonds, it is desirable that the government would reduce the corporate income tax (25%) and capital gains tax (5%) on the interest earned by domestic institutional investors through investing in green bonds. Considering green bonds' innovation varieties, government funds would provide guarantees for related bonds, realize external credit enhancement, reduce the financing costs of such bonds, and increase the enthusiasm of issuers. To guide third-party governance companies responsible for the pollution into issuing green bonds, part of the financial subsidy funds can be used as a discount measure for issuing bonds. In addition, it would also be encouraging if the government could enhance the environmental impact assessment dimension within the bond rating system or within the framework of existing rating companies.

In the existing information disclosure system, the corresponding disclosure of the use of raised green bond funds is tracked and disclosed largely in a voluntary manner. We would recommend that specialized asset appraisal institutions for special green assets gradually be cultivated and standardized so that when green asset securitization products are issued, they can achieve credit enhancement and reasonably reflect the value of green assets. These special green assets would include carbon assets and green income rights, such as new energy generation revenues. In case the green investment preferences of domestic investor groups need to be gradually cultivated, it is recommended that qualified foreign institutional investors (QFII) and qualified foreign institutional investors (RQFII) be introduced into the inter-bank bond market, through which they would drive the demonstration effect and gradually cultivate domestic investors' investment preferences for green bonds.

Additionally, collaboration between China and the international market in green bonds could be enhanced through pursuing genuine sustainable developments under the Belt and Road Initiative. More qualified domestic entities would be encouraged to participate in the development of international green bonds in collaboration with foreign institutions to leverage on the emerging green bond market of China.

6.6.2 Regional Policy—Greater Bay Area as an Example

A gradual integration of regional policies would complement the overall national policy for green financing. For example, the Greater Bay Area is at a point of rapid industrial transformation and growth, and the demand for green financing is substantial. The development of policy for green financing could inject new energy into the expansion of green business by financial institutions in the Greater Bay Area. Although a strategic combination of green finance and the Greater Bay Area policy is still in its infancy, such development prospects could be extensive given the vast potentials. Moreover, the Greater Bay Area is to present has various advantages and opportunities for project finance of green infrastructural developments.

All in all, the Greater Bay Area is an important, emerging economic development area; whereas, green finance is considered as a complementary financing mechanism for sustainable developments. The rapid development of the Greater Bay Area's green bond market would serve as an effective source of funding to fuel its sustainable investments and growth. Successful implementation of this concerted regional policy would inspire other regions to follow. Effective execution and implementation of such strategic regional development policies are in turn crucial to the development of the green debt market.

References

1. Cowan E (1999) Topical issues in environmental finance. Asia Branch of the Canadian International Development Agency (CIDA) (1):1–20
2. Jeucken M (2001) Sustainability in finance: banking on the planet. Earthscan, UK
3. Cowton CJ, Thompson P (2001) Financing the social economy: a case study of Triodos Bank. Int J Nonprofit Voluntary Sect Market 6(2):145–155
4. Ng AW (2018) From sustainability accounting to a green financing system: institutional legitimacy and market heterogeneity in a global financial centre. J Clean Prod 195:585–592
5. Wood D, Grace K (2011) A brief note on the global green bond market. In: Initiative for responsible investment working paper. Harvard University
6. Kaminker C, Stewart F (2012) The role of institutional investors in financing clean energy. In: OECD working papers on finance insurance and private pensions
7. Park A, Ren C (2001) Microfinance with Chinese characteristics. World Dev 29(1):39–62
8. Scholtens B, Cerin P, Hassel L (2008) Sustainable development and socially responsible finance and investing. Sustain Dev 16(3):137–140
9. Aizawa M, Chaofei (2010) Green credit, green stimulus, green revolution? China's mobilization of banks for environmental clean up. Environ Dev 19(2):119–144

10. Linnenluecke MK, Griffiths A (2010) Corporate sustainability and organizational culture. J World Bus 45(4):357–366
11. Climate Bonds Initiative and China Central Depository & Clearing Co., Ltd. (2018) China green bond market newsletter Q3. Climate bonds initiative. https://www.climatebonds.net/newsletter/china

Guangdong's Carbon Trading System: A Review of Liquidity and Influencing Factors

Jingyan Fu, Menghao Wu and Junyan Liu

Abstract China's seven carbon-trading pilots were approved for establishment in 2011 and subsequently commenced operations starting June 2014. As the largest carbon market among the seven Chinese carbon-trading pilots, Guangdong is evaluated as the second-largest carbon market in the world, only being next to the European Union (EU). These pilots instigated enormous influence on the policy formulated towards the establishment of China's national unified carbon market in 2017 with implications for solutions to environmental and climate problems on a national and even global level. This paper mainly adopts a comparative analysis approach, first to compare the liquidity of the Guangdong carbon market nowadays with a prior period and then with the Hubei carbon market. It further compares the influencing factors of carbon market liquidity based on the former results. The analysis identifies the factors influencing the low liquidity of the Guangdong carbon market, including the separation between the primary and secondary markets, insufficient openness in the secondary market, lack of adequate investors, inadequate consideration of the distribution and collection of emission allowances, an irrational carbon finance product structure, as well as a backward trading mechanism. Finally, this paper offers suggestions and proposals based on the problems identified in the overall assessment.

Keywords Guangdong Province · Hubei Province · Carbon market · Liquidity · Comparative analysis · Influencing factor

---

J. Fu (✉)
College of Economics, Institute of Resource, Environment and Sustainable Development, Jinan University, Guangzhou, China
e-mail: tfjyan@jnu.edu.cn

M. Wu
Personal Finance Department, Jiangsu Suning Bank Co., Ltd, Nanjing, China

J. Liu
College of Finance and Trade, Guangdong Industry Polytechnic, Guangzhou, Guangdong, China

J. Fu and A. W. Ng (eds.), *Sustainable Energy and Green Finance for a Low-carbon Economy*, https://doi.org/10.1007/978-3-030-35411-4_4

1 Introduction

To combat climate change, prior studies show that large-scale deployment of carbon capture and storage (CCS) is necessary to limit the increase in global average temperature to less than 2 °C by 2100 (Jorge H. García et al. 2018). Regarding the question of a carbon tax versus emissions trading, an ETS garners greater support than a carbon tax among stakeholders [1]. Many economists and policy-makers advocate price-based approaches, such as greenhouse gas emissions taxes and emissions trading programmes, or technology-based approaches, such as R&D subsidies and public–private R&D partnerships. The world generally seems to have become less (multi) polarized in terms of carbon emissions since 1992 [2]. Ultimately, however, both types of approaches rely on consumers and firms to make different choices [3].

Guangdong leads China's Provinces in industry, CO_2 emissions, and the size of its carbon-trading pilot. Understanding the driving forces of carbon emissions is necessary for policy formulation [4]. What changes will carbon emissions bring to China? The Regional Greenhouse Gas Initiative among the north-eastern states is expected to lead to an increase in the price of electricity in the region and beyond. In the RGGI region, the change depends on the portfolio of assets held by affected firms [5]. Industrial production and energy consumption have a significant positive impact on carbon emissions in both the short and long runs. Likewise, Granger causality analysis indicated a unidirectional causation from both industrial production and energy consumption to carbon emissions; e.g., the industrial and economic development in Bangladesh is taking place at the cost of environmental quality [6].

On December 19, 2013, Guangdong officially launched carbon emissions trading completing seven transactions on that first day; its total turnover is approximately 120 thousand tonnes and its average price is 60.167 yuan. During the first two years, Guangdong implemented a number of measurements to enliven the market, eventually occupying half of the nation. By July 27, 2015, the Guangdong carbon market's volume exceeded 20 million tonnes and the total turnover had reached 916 million yuan, accounting for 36.12% and 45.28%, respectively, of the entire nation's transactions. The Guangdong carbon market is generally good; it is not only at the vanguard of carbon trading and an important contributor to China's carbon market but also the third-largest carbon market in the world that is next to only the EU and Korea in terms of market scale.

However, the development of the Guangdong carbon market shows some worrying trends. From inception to the second commitment period, the liquidity of the market has demonstrated obvious deficiencies compared to those of Hubei, Shenzhen and the European Union. In the entire year of 2014, the Guangdong carbon market traded thousands tonnes of carbon allowance; the average price was 44.28 yuan per tonne, the highest price was 77 yuan per tonne and the lowest price was 21 yuan per tonne, showing great fluctuations in price. Compared with the Hubei market's annual volume of 7 million tonnes, the highest price of 26.59 yuan per tonne and the lowest price of 22 yuan per tonne, the Guangdong carbon market has the characteristics of smaller trade volume, dramatic price fluctuation and centralized transaction dates.

2 Literature Review

2.1 The Meanings of Carbon Market Liquidity

Because academics have not yet clearly defined liquidity in the carbon market, this paper will generally divide carbon market liquidity into four dimensions according to the above definition of liquidity:

The first dimension is breadth. Breadth is a concept that relates to price, generally exists in a carbon market with a market maker system and is often measured by quote spread. A small quote spread means the competition in the carbon market is more intense and the liquidity of the market is better. If there is no spread, the seller and buyer can reach a deal in accordance with their expectations, and the liquidity at this moment is the best. In one market with good breadth, the influence of transactions on the market is small.

The second dimension is depth. Depth is a concept relating to quantity and refers to the maximum volume traders can reach in a certain price. The greater the number is, the more depth the carbon market has, and the better liquidity. Gagelmann [7] asserts that if there are large amounts of transaction requests (orders) within a certain scope of the current price, then the carbon market has more depth. A carbon market with great depth means its liquidity is good, and vice versa; the same transaction may produce greater effects on the price in a low-depth market.

The third dimension is immediacy. Immediacy is a concept regarding time, referring to the time that a trader must wait to conduct a transaction. If an investor can deal more quickly at a certain price, then the liquidity of the carbon market is better. Prohibiting the intertemporal trading of emission allowances induce positive risk premium in futures prices when the trading of the contracts and their expiry take place in time periods separated by this trading ban [8].

The fourth dimension is resiliency. This means the speed at which prices return to the state of equilibrium after certain transactions cause fluctuations in price. If prices can return to equilibrium more quickly, the capacity of the carbon market to bear transactions is greater, its resiliency is higher, and its liquidity is better. In a highly flexible carbon market, it is difficult to produce unbalanced price.

2.2 General Influencing Factors of Carbon Market Liquidity

The Bank of International Settlements (BIS) noted in a report (1999) that the influencing factors of market liquidity can be divided into three categories: the design of the product, the market microstructure and the behaviour of the market participants. According to the BIS classification, this paper divides the influencing factors of the carbon market into three types:

Design of product

Designing a product that can match the diverse needs of different market participants is the first factor that should be taken into account. Only when we design a product to meet the needs of the majority can the enthusiasm of the market participants be increased, and as a result, the market liquidity may be improved.

Market microstructure

The microstructure of the market mainly includes the trading mechanism, market participants, transaction cost and transaction constraints. The trading mechanism is channelled for the sides to trade, and the common transaction mechanism includes the following three types. First, the buyer and the seller transact directly without any intermediate process. The second type is the market maker system, in which the sides must deal through the market maker, and the market maker quotes the respective price to each side. The third type is pairing and dealing automatically by a system, such as the call auction and the continuous auction.

Behaviour of market participants

Different characteristics of the participant will affect its decision-making and behaviours, such as its ability to bear risk, sensitivity to macroeconomic policies and expectations about the future market. When risk-averse investors increase, market liquidity may decrease. When the expectations of the future of all of the investors are the same, the market liquidity will also decrease. Therefore, the diversity of market participants is good for market liquidity. Additionally, laws, regulations and macroeconomic policy will influence the participants' behaviour.

2.3 Measurements of Carbon Market Liquidity

According to the theory of financial market microstructure, market liquidity is multidimensional. Fischer [9] proposed that the measurement of liquidity must factor in four conditions: depth, width, tightness and resilience. Although scholars have performed considerable research on how to measure market liquidity, there is still no consensus.

Current methods of measuring market liquidity focus mainly on four different perspectives: methods related to bid-ask spread, methods related to the influence of trading volume on price, methods combining price and volume and methods related to the time consumed waiting for trade. Among the above four, the methods related to bid-ask spread are mainly suitable for the quote-oriented market; it is not certain whether these methods suit the order-oriented market and the carbon market is not a market based on the market maker; as a result, these types of methods are not the most suitable for measuring carbon market liquidity. Additionally, the methods related to the influence of transaction on price include the market depth model of Kyle [10] and the simple transaction model; these methods often need great quantities of data on the micro-market in the actual analysis, and the methods related to the time consumed waiting for trade are very much the same.

Because this paper must measure the daily liquidity of the carbon market, after considering the availability of data and the differences among the carbon markets, this paper opts for the methods combining the price and trade volume. Taking account of the suitability of each liquidity ratio, this paper ultimately uses a new method that combines the Martin index [11] and the Hui and Heubel liquidity ratio [12]; the formula is as follows:

$$\mathrm{MH} = \frac{|(P_t - P_{t-1})/P_{t-1}|}{V/Q_{\mathrm{ea}}} \tag{1}$$

In the formula, "MH" is the new liquidity index, "V" represents the trading volume of the carbon emission allowance on the exact trading day and "Q_{ea}" represents the amount of carbon allowance in circulation in the entire carbon market and is equal to the total amount of the designed allowance minus the allowance that has not been distributed. "V/Q_{ea}" represents the turnover rate of allowance. "P_t" represents the closing price on "t" trading day, and the numerator of the formula is the volatility of price on "t" trading day.

3 Liquidity Analysis of the Guangdong Carbon Market

3.1 Guangdong Carbon Market Liquidity

When there are no transactions or the price is completely stable, the MH index is not affected to reflect the fact of liquidity itself.[1] The trend of zero transactions, the liquidity with non-fluctuating price and the liquidity on a normal trading day are analyzed in the following.

3.1.1 The Trend of Zero Transactions

From December 19, 2013 to November 13, 2015, there were a total of 471 trading days (excluding holidays and temporary adjusted trading days); on 38.34% of those days, equal to 181 trading days, there were zero transactions. The monthly occurrence of zero transactions during this period is shown in Fig. 1.

The monthly frequency of zero transactions in the Guangdong carbon market showed a downward trend, and the market performance in the year 2014 was much better than that in 2013, especially the highest proportion, which was only 60% in 2014 and now reaches 100%. Second, the monthly proportion of zero transactions

[1] When there are no transactions, the MH index is invalid. When the price has no fluctuation, the MH index always remains at zero.

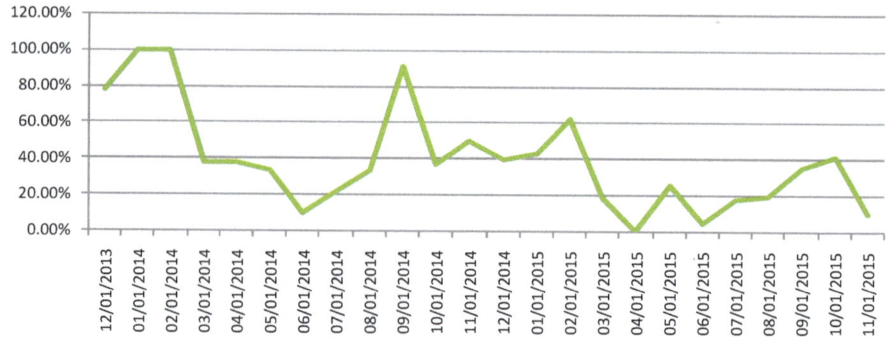

Fig. 1 Trend of monthly zero transactions from 2013/12/19 to 2015/11/13 in CEEX. *Source* http://www.cnemission.cn

in 2014 increased from 20 to 40% while it grew from 0 to 100% in 2013, indicating that the Guangdong carbon market was extremely unstable in 2013 and improved somewhat in 2014. Individually, there were obvious "V" cycle characteristics in both years. The proportion in June was the minimum of each year, and the market performance was the best.

3.1.2 The Liquidity with Non-fluctuating Price

There were a total of 23 days on which transaction did not affect the price from December 19, 2013 to November 13, 2015 (having ruled out situations where there were no transactions on a trading day), nine of which were in 2014 and the rest in 2015 (details are shown in Tables 1, 2).

Table 1 Dates and quantities of GDEA flowing into the market

Trading date	Trading volume (tonnes)	Trading date	Trading volume (tonnes)	Trading date	Trading volume (tonnes)
2014.03.11	5242	2014.11.13	2000	2015.07.28	1600
2014.03.18	100	2015.01.19	500	2015.08.14	25,000
2014.04.16	6	2015.01.22	1913	2015.08.17	26,833
2014.10.16	300	2015.01.26	1913	2015.10.21	1661
2014.10.30	1	2015.01.28	3	2015.10.22	58,539
2014.11.05	1000	2015.03.03	500	2015.10.23	20,500
2014.11.06	1000	2015.03.06	3000	2015.11.10	300
2014.11.07	1	2015.03.13	2000		

Source http://www.cnemission.cn

Table 2 Relevant regulations and policies of the Guangdong carbon market in 2015

Country level	"Overall plan for the reform of ecological civilization system"	September 21, 2015
	"USA-China Joint Presidential Statement on Climate Change"	September 25, 2015
	"France and China agree to monitor climate change pledges"	November 2, 2015
Province level	"The 2015 implementation plan of Guangdong carbon allowance"	August 18, 2015
	"The implementation plan of greenhouse gas emission report on key enterprises of Guangdong Province"	September 21, 2015
	"Notice on carrying out management of carbon emissions on reporting historical carbon emission information in Guangdong Province"	September 23, 2015
Exchange level	"Trading rules for carbon allowance"	August 31, 2015
	"Guidelines for carbon allowance repurchase business in Guangdong"	October 21, 2015
	"The process of registration and operation on carbon allowance mortgage business in Guangdong Province (for Trial Implementation)" "Detailed rules for controlling and managing carbon trading risk"	December 17, 2015

Source http://www.cnemission.cn

In terms of the trading dates shown in Table 2, the number of trading days without any price fluctuation continued increasing from 2014 to 2015. There were 7 such days during the first four months of compliance year 2015, which indicates the market was becoming increasingly stable. However, the liquidity still cannot be clearly determined under the circumstance of no price fluctuation on the trading day. Therefore, it is necessary to compare the trading volumes of the trading dates in Table 2; the trading volume trend of the 23 days is shown in Fig. 2.

Table 2 shows that the trading volume experienced a dramatic increase on August 14, 2015, and the overall trading in the following period was better than in the early stages. This shows that the trading volume's influence on price was gradually weakening and the ability of the Guangdong carbon market to bear risk had improved; the Guangdong carbon market was quietly changing.

3.1.3 The Liquidity on a Normal Trading Day

Further, we can directly use formula (1) to measure the liquidity of the Guangdong carbon market, and after eliminating several extreme values, which may disturb observation, this paper arrives at a figure to reflect the liquidity trend of a normal trading day on the Guangdong carbon market as shown in Fig. 2. We first see from Fig. 2 that there is a difference between the years 2014 and 2015; the MH index of

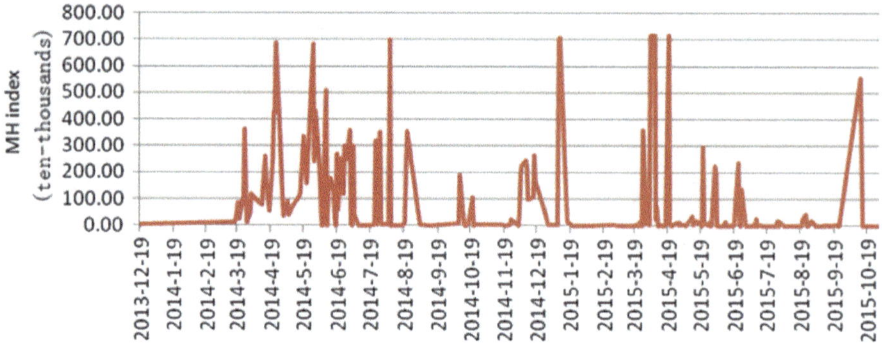

Fig. 2 Guangdong carbon market's MH index on a normal trading day. *Source:* http://www.cnemission.cn

the former was more centralized, and the majority of it basically remained above 100 million, while the MH index of the latter was scattered and had only a few high values, many of which were almost approaching zero.

3.2 *Influencing Factors of the Liquidity of the Guangdong Carbon Market*

Due to differences in liquidity in periods around compliance year 2015 in the Guangdong carbon market, to investigate why the liquidity changed and identify the factors that influenced this liquidity, this section will divide the development of the Guangdong carbon market into two periods: compliance year 2014 (from July 16, 2014 to June 23, 2015) and compliance year 2015. It will then compare the two stages from four aspects in accordance with the classification of the general influencing factors of the carbon market.

3.2.1 Laws, Regulations and Policies

Policies and regulations generally remain effective in the long term once promulgated; for instance, the National Development and Reform Commission of China published the "Interim Measures for the Management of Carbon Emissions Trading" on December 12, 2014, to take effect 30 days later, but there is no specific period of validity, which means that the measures will be implemented over the long term. The Guangdong Development and Reform Commission officially implemented "the Detailed Rules of Managing Carbon Emission Allowance" on March 1, 2015 and stipulated that the rule would be valid for 5 years. The regulations and policies implemented in compliance year 2014 were effective in compliance year 2015. To determine changes in regulations and policies in 2015, here, comparison of the regulations and policies in the two stages is limited to analyzing the new regulations and policies published in compliance year 2015. The new policies and regulations related to the Guangdong carbon market are shown in Table 2.

As Table 2 shows, at the country level, there were a total of three policies related to the Guangdong carbon market in compliance year 2015, at the provincial level four, and at the exchange level three. At the country level, China has put forward comprehensive plans for ecological civilization construction, which include the contents of the carbon rights system and green finance, and has reached consensus with the USA and France, demonstrating a determination to actively participate in tackling climate change. At the provincial level, the Guangdong Provincial Development and Reform Commission made detailed provisions regarding the issuance of Guangdong carbon allowances for the 2015 compliance year and officially launched reporting of greenhouse gas emissions. At the exchange level, CEEX further optimized their trading rules and strengthened their ability to control trading risk, and in the meantime also opening up the business of buying back and mortgaging carbon allowances.

Comparing regulatory policies, we see that there were no major changes between the compliance years 2014 and 2015. The state released two documents about solving climate problems, which indicate that China is gradually paying more attention to ecological and climate change. The prospects for China's carbon market development remain bright, but in the compliance year 2015 there were no major influential laws or regulations released. Additionally, except for the four rules published by CEEX, no important and relevant laws or regulations were published at the provincial level; the only major difference was the "carbon allowance scheme", which may be one reason why the liquidity of the Guangdong carbon market changed in 2015. In addition, we can determine that Guangdong's carbon financing business has begun to improve and become standardized.

3.2.2 Carbon Allowance Scheme

Compliance year 2014

The total carbon allowance of Guangdong in 2014 was approximately 0.408 billion tonnes, including 0.37 billion tonnes for emission-controlled enterprises and 0.038 billion tonnes for reserve. The reserved allowances consisted of allowances for new business projects and allowances for market adjustments. There were in total 193 emission-controlled enterprises from the industries of power, cement, petrochemicals and steel, whose annual emissions exceeded 20 thousand tonnes of CO_2 (or comprehensive energy consumption equal to 10 thousand tonnes of standard coal). Additionally, 18 new business projects had the same emission standard as the emission-controlled enterprises after construction and operation. The allocation method of carbon allowances combined the baseline method and the historical method (Table 3).

Table 3 Calculation methods for carbon allowances of Guangdong's enterprises in 2014

	Baseline method		Historical emission method	
Applied enterprise	Electric power industry coal-fired gas generating set + Production and grinding of ordinary cement clinker in cement industry enterprise with long process in iron and steel industry		The other emission-controlled enterprises	
Calculation formula	Emission-controlled enterprise	Advance allowance = actual production in 2013 × benchmark value × annual decline coefficient (1)	Emission-controlled enterprise	Allowance = historical average annual emissions × annual decline coefficient + process flows allowance (for upgrading of oil products in petrochemical enterprises)
		Certificated allowance = advance allowance × production correction factor		
	New project/enterprise	Allowance = designed capacity × benchmark value	New project/enterprise	Allowance = estimated annual comprehensive energy consumption × carbon emission reduction factor

Source http://www.cnemission.cn

Because of the special provisions regarding the output correction factor of coal-fired generators and gas generators, the actual allowance for the power industry is relatively less, and its need to expand capacity is somewhat limited.

Compliance year 2015

The total carbon allowance of Guangdong in 2015 was also approximately 0.408 billion tonnes consisting of 0.37 billion tonnes for emission-controlled enterprises and 0.038 billion tonnes for reserve (reserved allowances included allowances for new projects and market adjustments). The emission-control system covered 186 enterprises, whose emissions exceeded 20 thousand tonnes (or comprehensive energy consumption equal to 10 thousand tonnes of standard coal) and came from the four industries of power, steel, petrochemicals and cement. There were 31 enterprises with new projects, and the standard to be covered was the same. The allocation method of allowances adopted the baseline method and the historical method (details are shown in Table 4).

The specific allowance calculation methods for 2015 were almost the same as in 2014, and there was no special provision regarding the output adjusting factor of some of the industries and related processes. The allocation of allowances still combined the free and the paid, with the free allocation proportion of the power industry at 95% and for the other three industries at 97%.

Comparison

Comparing the Guangdong carbon market's allowance allocation system in compliance years 2014 and 2015, we see that the allowance allocations of the two years were generally similar but that there were also many differences in the details. Summaries of several major changes in Guangdong's carbon allowance allocation in compliance year 2015 are presented in Table 5.

3.2.3 Trading Rules

The Guangdong carbon market consists of the primary market and the secondary market. The annual trading scale of the primary market is significantly higher than that of the secondary market, but taking into account the few trading opportunities and large time intervals, the comparison here focuses mainly on the secondary market in compliance years 2014 and 2015.

(1) The common points of the two compliance years

The participants in the Guangdong carbon market in both years included emission-controlled enterprises, new projects, individuals, some approved investment institutions and other organizations. The market implemented a membership management system, and its products were GDEA and CCER. The transactions model adopted listing-order selection and transfer by agreement (the detailed process is shown in Fig. 3). The single order to list allowances above 100 thousand tonnes was required

Table 4 Calculation methods for carbon allowances of Guangdong's enterprises in 2015

	Baseline method		Historical emission method	
Applied enterprise	Electric power industry coal fired gas generating set + Coal-fired hot spot co production unit production and grinding of ordinal cement clinker in cement industry enterprise with long process in iron and steel industry		The other emission-controlled enterprise	
Calculation formula	Emission-controlled enterprise	Advance allowance = actual production in 2014 × benchmark value × annual decline coefficient (1)	Emission-controlled enterprise	Allowane = historical average annual emissions × annual decline coefficient
		Certificated allowance = advance allowance × production correction factor		
	New project/enterprise	Allowance = designed capacity × benchmark value	New project/enterprise	Allowance = estimated annual comprehensive energy consumption × carbon emission reduction factor

Source http://www.cnemission.cn

Table 5 Changes in GDEA's allocation mechanism in compliance year 2015

Change 1	Change 2	Change 3
Emission-controlled enterprises increased from 2011 to 2017	Coal-fired cogeneration unit was brought into the scope of applying the baseline method	The amount of paid allowance was reduced from 8 million tonnes to 2 million tonnes
Change 4	**Change 5**	**Change 6**
Bidding of paid allowance implemented the policy reserved price to replace the auction base price	The restriction that enterprises could only buy carbon allowances from the primary market was eliminated	The process flow in some industries was eliminated

Source http://www.cnemission.cn

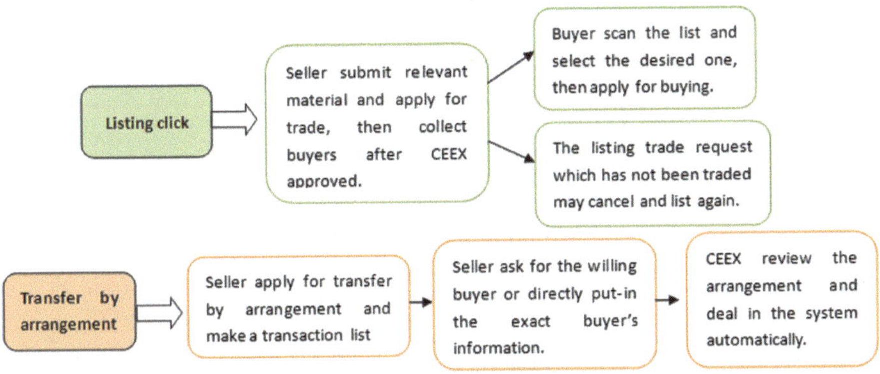

Fig. 3 Process of the listing-order selection model and the transfer by arrangement model. *Source* http://www.cnemission.cn

to use the model of transfer by arrangement, which differs from the listing-order selection model in that with less information to list, it previously had not needed to include the desired price and quantity. The opening price is the closing price of the last trading day, and price fluctuation is limited to 10%.

(2) The adjustments to trading rules in compliance year 2015

The 2015 transaction fee was changed significantly, and CEEX made other adjustments in the trading rules. On June 8, 2015, CEEX formally implemented many preferential measures for carbon trading, lowered its threshold and put forward features to attract and encourage investors to trade (shown in Table 6).

In comparison, there were many substantial changes to the trading aspect. The change in the calculation method of the closing price may have resulted in a more reasonable closing price to reflect the real market price and avoid manipulation risk. The transaction fee also had to be reduced dramatically with 60% less than

Table 6 CEEX trading rule adjustments comparison in 2015

	Closing price	Transaction fee	Transaction preferential	Account opening process	Listing and selection	Monitoring account	Membership preferential
The compliance year of 2014	Weighted average price of the last ten transactions	Charge 5‰ of the transaction amount on both sides	\	Open account on site with materials	1. Disorder list 2. Can only choose the lowest sell order and the highest buy order	\	\

(continued)

Table 6 (continued)

	Closing price	Transaction fee	Transaction preferential	Account opening process	Listing and selection	Monitoring account	Membership preferential
The compliance year of 2015	Weighted average price of all transactions in one day	Charge 2‰ of the transaction amount on both sides	Individual member: If traded quantity accumulated up to 3000 tons, the next year's annual fee can be exempted	Open account remotely by mailing materials	1 List can be sorted automatically by time and price 2. Can only choose the lowest sell order and the highest buy order 3. Can avoid "wrong selection" that resulted from operational errors	Can apply for one account, and check the transaction record and instant quotes if in need	If joined in before June 30th, 2016, all kind's member can be exempted the membership fee and 50% annual fee
			Institution member: If traded quantity accumulated up to 100 thousand, 50% of the next year's annual fee can be exempted, and a traing place worth 7000 yuan can be presented, too If traded quantity accumulated up to 200 thousands tons, the next year's annual fee is free, and 2 training places can be present				

Source http://www.cnemission.cn

previously, which may have been very attractive to investors. In addition, many preferential measures for the Guangdong carbon market were implemented in 2015.

4 Comparison of the Market Liquidity of Guangdong's and Hubei's Carbon Pilots

The Hubei carbon market is a relatively special carbon market among the seven Chinese carbon pilots; it started very late, on April 2, 2014. Hubei has the most stable trading and price. As the Carbon K-line indicates, the price of the Hubei carbon market generally remained at approximately 25 yuan per tonne from when it began to November 30, 2015. The trading volume is also stable, and it was affected relatively little by the time span. Additionally, there were almost no trading days without any trading.

4.1 Trends in the Liquidity of the Hubei Carbon Market

Hubei's carbon allowance allocation method is more complicated than that of Guangdong, especially in the data and information, except for the CHEEX's announcement that Hubei only conducted one transfer of 2 million tonnes government reserved allowance in 2014 and there was no transfer of government reserved allowance in 2015. It is difficult to access the other data and information to calculate the MH index of the Hubei carbon market. As a result, this paper will scale up the index and substitute the market liquid allowance with the entire year's total allowance in 2014, representing the market liquid allowance with 1.1 times the entire year's total allowance in 2015. In this way, the calculated market liquidity may appear worse than in reality.

(a) **The trend of zero transactions**

On the market from April 2, 2014 to November 13, 2015, there were 390 trading days (excluding legal holidays and the temporary announced market closing day). There were no trading days without any transactions; its proportion of zero transactions in each month was zero per cent.

In the market's simple zero transactions trend, Hubei's carbon market was very active, with trading days with more or fewer transactions accounting for 100% of the total working days. There were no trading days without any transactions, which is in sharp contrast to Guangdong's carbon market. Hubei's carbon market performed well from this point of view, and the market was stable and had adequate liquidity.

(b) **The liquidity with non-fluctuating price**

During the period from April 2, 2014 to November 13, 2015, there were a total of 23 trading days as shown in Table 7. Hubei's carbon market was the same as

Table 7 Dates and quantities of HBEA flowing into the market

Trading date	Trading volume (tonne)	Trading date	Trading volume (tonne)	Trading date	Trading volume (tonne)
2014.05.25	43,977	2015.01.23	20,528	2015.06.02	121,856
2014.08.08	16,181	2015.01.28	13,608	2015.06.18	19,807
2014.10.31	18,275	2015.01.30	66,828	2015.06.19	660
2014.11.05	7639	2015.02.06	14,690	2015.06.24	17,500
2014.12.30	32,000	2015.03.27	25,337	2015.06.25	16,000
2015.01.08	5675	2015.04.07	57,100	2015.06.30	12,095
2015.01.14	8207	2015.04.23	15,795	2015.07.30	68,171
2015.01.21	27,029	2015.04.24	47,755		

Source http://www.cnemission.cn

Guangdong's in total trading days with non-fluctuating prices. However, the Guangdong carbon market started approximately 5 months before Hubei's, which means the Hubei carbon market had more trading days with non-fluctuating prices than Guangdong's in the same period and that Hubei's carbon price performed better than Guangdong's in price stability. In addition to the differences in price and trading days, the trading volume of Hubei's carbon market was more stable. Except for the minimum 660-tonne allowance, the trading volume of the other trading days all exceeded 5 thousand tonnes.

The trading volume of the Hubei carbon market's trading days with non-fluctuating price essentially remained at approximately 20 thousand tonnes; the maximum exceeded 120 thousand tonnes on June 2, 2015. Hubei's compliance period of the first compliance year was from July 2, 2015 to July 10, 2015; however, the market had no obvious "compliance phenomena" in the period before compliance. Therefore, Hubei's carbon market was generally stable and without periodic features.

(c) **The liquidity on a normal trading day**

As in the analysis of Guangdong's carbon market, after excluding the above two situations, the remaining trading days were all normal in terms of trading volume and price fluctuation. As a result, here, formula (1) can be used to measure the Hubei carbon market's liquidity. After eliminating the two extreme values of 3503.741 on November 12, 2015 and 21.5796 on March 4, 2015, the MH liquidity index trend of Hubei's carbon market can be shown as in Fig. 4.

In these circumstances, although the MH index has been scaled up compared with the actual values, the overall MH values remain very low. Except for the several extreme values, the maximum is 4, and after eliminating this value, most values are close to the "X-axis". In addition, from the MH index trend, we see no obvious regularities and that the index values from August to September slightly increased. The index values of early November 2014, late February 2015 and late August 2015 all have the trend of drastic inclines, but because there are too few of these values,

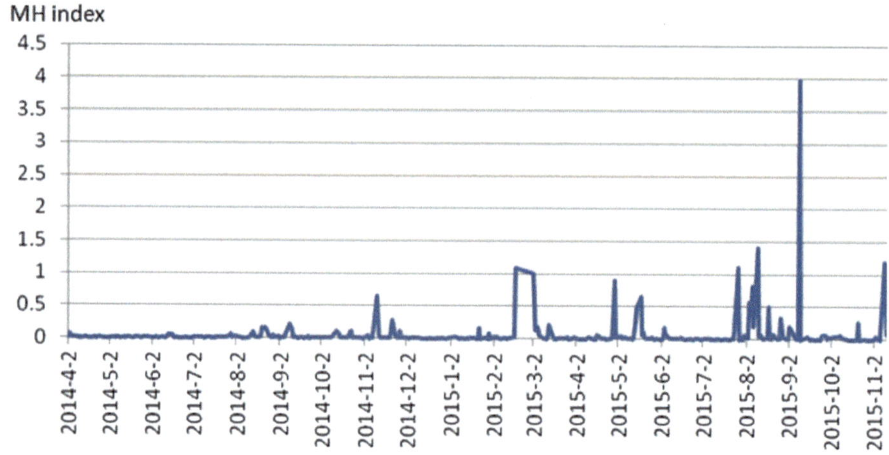

Fig. 4 MH Index trend of Hubei's carbon market. *Source* http://www.cnemission.cn

they are not quite representative. Therefore, the MH index values of Hubei's carbon market were relatively small and stable, which means the liquidity of Hubei's carbon market was adequate and stable.

4.2 Liquidity Comparison of the Guangdong and Hubei Carbon Markets

In general, Guangdong's carbon market was unstable and often had zero transaction trading days; its features included serious "compliance phenomena", great price fluctuation, and low exchange rate. After summarizing the liquidity analysis of the Guangdong carbon market from three different angles, we find that its overall liquidity was poor and there remained a large gap between it and Hubei's carbon market.

(1) Regarding the market's zero transaction days, Guangdong had a high proportion of such trading days each month, and the "zero transaction" proportion rose to 100%.
(2) For the liquidity with non-fluctuating price, Guangdong was the same as Hubei but owing to the late start of Hubei's carbon market, Guangdong's still lagged behind Hubei's in market stability.
(3) Except for the above two circumstances, on normal trading days, even when Hubei's MH index values were scaled up, they were still lower than Guangdong's. Guangdong's MH index values were too high, at almost ten thousand or one million times Hubei's MH index.

4.3 Comparing Influencing Liquidity Factors Between the Two Carbon Markets

4.3.1 Laws, Regulations and Policies

Relevant regulations and policies for the Hubei carbon market

According to the policies, laws and regulations listed in Hubei's carbon emissions trading centre bulletin, most of the policies and regulations are at the provincial level and only two were promulgated by the Hubei carbon emission exchange. This may have resulted from the provincial government's heavy dominance of Hubei's carbon market and the relatively strong attachment of Hubei's carbon emission exchange to the provincial government. Policies and regulations applying to the Hubei carbon market are shown in Table 8.

On the provincial level, Hubei put forward general requirements about the distribution and management of emission allowances, market trading, MRV systems and the incentive-constraint mechanism, while the allowance allocation method was also described in detail. In addition, the "Low-Carbon Development Plan of Hubei's

Table 8 Laws, regulations and policies of the Hubei carbon market

Province level	"Interim Measures for the administration and trading of carbon emission rights in Hubei Province"	April 4, 2014
	"The scheme of carbon allowance in Hubei Province"	April 14, 2014
	"Guidelines for the verification of greenhouse gas emission in Hubei Province (for Trial Implementation)"	July 24, 2014
	"Guidelines for the detection, quantification and reporting of greenhouse gas emissions from industrial enterprises in Hubei Province (for Trial Implementation)"	
	"Hubei Province low-carbon development plan of energy saving and emission reduction in 2014–2015"	October 27, 2014
	"Notice of the provincial development and Reform Commission on the relevant matters concerning the carbon emission right offset mechanism in Hubei Province in 2015"	April 17, 2015
	"Measures for the administration of Hubei Province on carbon emission allowance release and repurchase (for Trial Implementation)"	September 28, 2015
Exchange level	"Rules for the implementation of the carbon allowance custody business in Hubei Province (for Trial Implementation)"	December 8, 2014
	"Trading rules for carbon allowance in Hubei Province"	December 11, 2014

Source http://www.cnemission.cn

Energy saving and Emission Reduction" set a goal of 3.4% reduction in carbon emissions per GDP.

Comparison with Guangdong

In terms of carbon emission reduction targets, the "Guangdong Provincial Plan of Energy Saving and Carbon reduction for 2014–2015" set a reduction goal of more than 3.5% for carbon emission per GDP, which is basically consistent with Hubei Province. Guangdong also formally implemented the provincial regulation "Pilot Measures for the Administration of Carbon Emission in Guangdong Province" on March 1, 2014; considering the different starting dates of the two markets, this was relatively late for Guangdong to publish provincial carbon regulations. Additionally, the one major difference between the two markets is the disposal method of residual carbon allowances. Hubei provides that a residual carbon allowance that has not been traded must be cancelled in the compliance period and cannot be held by an enterprise itself, while Guangdong allows enterprises to withhold a residual carbon allowance and use it for the following year's compliance.

4.3.2 Carbon Allowance Scheme

Hubei Province

Hubei had a total of 0.324 billion tonnes carbon allowance in compliance year 2014, which consisted of the initial allocation allowance, the reserved allowance for adjustment and the government's reserved allowance. The concrete calculating methods are as follows:

- The initial allowance = total carbon emission of enterprises in 2010 × 97%
- The government's reserved allowance = total carbon allowance × 8% (includes 30% for public bid)
- The reserved allowance for adjustment = total carbon allowance − the initial allowance − government's reserved allowance

The initial allowance is freely issued to enterprises at one time early in the compliance year, while the reserved allowance for adjustment is issued after enterprises complete their emission verification before compliance. The government's reserved allowance is mainly used to adjust the supply and demand of allowances and further maintain market stability. On March 31, 2014, the 2 million tonnes of Hubei government's reserved allowance was successfully bid, and the base price was 20 yuan per tonne allowance.

If the actual carbon emission and the initial allowance differ more than 20% or 2 million tonnes, enterprises are allowed to apply to change their initial allowance (the detailed calculation methods are shown in Tables 9 and 10).

Table 9 Carbon allowance calculation method of Hubei's emission-controlled enterprises in 2014

Industry	Calculation formula
Electric power industry (baseline method)	Total amount of carbon allowance = advance allowance ± post-regulation allowance Advance allowance = historical base emission × adjustment coefficient 0.9192 × 50% Post-regulation allowance can be divided into additional allowance and collected allowance Additional allowance = excess generating capacity × benchmark value (99.193 tonnes/thousand kilowatt hours for thermal power, combined heat and power, coal gauge power generation enterprises adopt the emission of unit power generation in the exact year) Collected allowance = residual electricity generation × carbon emissions of unit power generation
Non-electric power industry (historical method)	Initial allowance of the compliance year = historical base emission × adjustment coefficient 0.9192

Source http://www.cnemission.cn

Table 10 Calculation method of adjusted allowance of Hubei's emission-controlled enterprises in 2014

Difference of more than 20% between actual carbon emissions and the annual initial allowance:
Additional allowance = actual carbon emission-initial allowance × 120% Collected allowance = initial allowance × 80%-actual carbon emission
Difference of more than 200 thousand tonnes between actual carbon emissions and the annual initial allowance:
Additional allowance = actual carbon emission-initial allowance-200 thousand tonnes Collect allowance = initial allowance-actual carbon emission-200 thousand tonnes

Source http://www.cnemission.cn

Comparison with Guangdong

In terms of setting the total carbon allowance, Guangdong Province allows more than that of Hubei. The Guangdong government's reserved allowance accounts for 9.3% of the total allowance, 1.3% higher than that of Hubei, and the allowance used for public bidding is also more than that in Hubei. Additionally, both Guangdong and Hubei adopt the allocation method of rolling the base three years but taking account of the different industrial structures and Hubei's rapidly growing economy, Hubei's carbon allowance is relatively tight. One significant reason for the heightened activity in the Hubei carbon market is its tight carbon allowance; the calculated allowance in accordance with the base year is obviously inadequate for the economic reality.

4.3.3 Trading Rules

Hubei's carbon market consists of a primary market and a secondary market. However, the secondary market is considered as the main component, whereas there is only one auction in the primary market. The following focuses mainly on analyzing the secondary market.

Hubei's carbon trading rules

Hubei's primary market also chose auction as the only transaction mode; however, the allowance that has been traded comes from the government's reserved allowance. In terms of the market openness, Hubei's primary market is opened to enterprises, investment institutions, and social and individual investors. In terms of the auction price, Hubei set a price as the base trading price in compliance year 2014, which is the same as Guangdong. The base price "20 yuan per tonne" is very close to the secondary market. However, Hubei only conducted one auction in compliance year 2014.

For the secondary market, the participants include domestic and foreign investment institutions, carbon emission-controlled enterprises, organizations and individuals, all managed by the membership management system of CHEEX (see Table 11). Until October 30, 2015, the Hubei carbon market had a total of 6,292 investors. The market uses the mixed transaction mode of "pricing transfer" and "negotiated transfer", and all transactions must declare their trading demand through the trading system. In the meantime, the allowance quantity that individuals hold cannot exceed 1 million tonnes.

Table 11 Transaction mode of Hubei's secondary carbon market

Negotiated transfer	Pricing transfer	
"Negotiated transfer" implemented the non-continuous trading forms; 5 min is a complete period, the previous 4 min is for declaration and the last 1 min is for negotiation and report. The price can fluctuate in the range of 10% above and below of the last trading day's closing price	Public transfer	Transfer by arrangement
	1. Fixed price 2. List first and then wait for transaction order 3. According to the principle of price priority	1. The same as Guangdong's transfer by agreement 2. Take the same 10% price fluctuation range with public transfer

Source http://www.cnemission.cn

The opening price of Hubei's secondary carbon market is the price of the first transaction, and the closing price is equal to the weighted average of the prices in the last 10 trading periods. If there has been no transaction in the last 10 trading periods, the closing price is represented by the last transaction of the trading day. In terms of the transaction costs, individual and institutional investors are free to open an account, broker members who joined before September 30, 2015 are exempt from the membership and annual fees, and the brokerage commission ratio will gradually increase from 50% to 90% along with its transaction size. Additionally, there are two notable features of the handling fee. First, the handling fee of negotiation and bargaining does not exceed 5% of the total trading amount on both sides. If the total handling fee reaches 100 thousand yuan or one must pay a handling fee of 100 thousand yuan at one time, investors can request a fee waiver. Second, the handling fee of a pricing transfer is 4% of the total trading amount on the single seller.

Comparison with Guangdong

First, the source of allowance for auction in Hubei differs from that in Guangdong's for auction comes from the allowance of enterprises which carbon emission is controlled. For the primary market in Guangdong, individual investors are not allowed to participate, and the conditions of institutional investors entering the primary market are somewhat strictly regulated.

Second, based purely on Hubei's carbon market, although there is only one auction in the primary market, the secondary market liquidity has been very stable and adequate. Therefore, the primary market may have limited influence on market liquidity.

Third, the auction price of Hubei's primary market is close to that of the secondary market, which is similar to that of Guangdong after Guangdong established the link between the primary market price and the secondary price. We could infer that the primary market price may have greatly influenced the carbon market liquidity.

Fourth, participants in the Guangdong and Hubei secondary carbon markets are essentially similar, including enterprises, institutional investors and individual investors.

Fifth, the daily price limits in Guangdong and Hubei are basically consistent; the price limit of "negotiated transfer" in the Hubei carbon market is, as in Guangdong, 10%. The price limit of "pricing transfer" is different at 30%. Hubei has not restricted institutional investors buying allowances; the only restriction is that individual investors must hold allowances of less than 1 million tonnes. However, Guangdong has placed restrictions on their institution investors and individuals, with the maximum holding allowance 3 million tonnes.

Sixth, in terms of the transaction costs, Hubei has lower costs than Guangdong. Hubei has not only allowed institutions and individuals to open accounts freely but also implemented relief measures on account, annual and handling fees. The gradient commission discount, in particular, is more conducive to encouraging members to trade and further stimulated its carbon market.

4.3.4 Carbon Finance

While Guangdong's carbon finance is more focused on enterprises' compliance and financing, Hubei's carbon finance is more diverse and with a variety of functions (see Table 12). Regarding carbon funds, which can effectively drive market trading, Guangdong also differs from Hubei. Guangdong's carbon fund is government-oriented and with low marketization. Private capital has not fully attracted participation in carbon trading, which may be related to the different positioning of the two carbon markets.

In terms of time, the progress of developing carbon finance in Guangdong Province is relatively slow compared with that in Hubei Province. Guangdong has launched only four carbon financial products in the 2 years since it started. Although Guangdong established a carbon fund, it is still too dependent on the provincial government to promote the needed degree of marketization and operate it successfully. Hubei's carbon finance has been blossoming everywhere from the beginning of the market, achieving successful operation of the carbon emission right pledge loan, carbon fund, carbon asset custody, carbon financial credit and carbon bond in only several months.

In terms of the scale of products, Hubei has promoted many carbon emission right pledge loans for a total of 0.54 billion yuan over one year, while Guangdong has just finished once for a total of 5 million yuan. The pledge loan activity in Guangdong is still far from that in Hubei. For the other three financial products of Guangdong's carbon market, no corporation has used the carbon trading corporation overdraft since it was launched. As the new carbon financial products of Guangdong, the carbon allowance repurchase and EA-SCP are very important mechanisms.

5 Conclusions

Comparing its own performance over time, the liquidity of Guangdong's carbon market has had obvious compliance phenomena in the previously discussed 2 compliance years. Liquidity was relatively better during the period around compliance, but on normal trading days it was extremely poor. In compliance year 2015, there were signals that the market liquidity might change and improve. But compared with Hubei Province, owing to the obvious inadequate liquidity, there remained much room for Guangdong to improve its carbon market liquidity. Through the comparison of market liquidity and its influencing factors from historical and lateral perspectives, this paper draws the following conclusions:

First, because the Guangdong carbon market did not publish many policies and regulations in compliance year 2015, the liquidity was contrary to the usual trend and performed successively better in the months after the last compliance period. This change indicates that there is no great relevance of policies and regulations to the liquidity of Guangdong's carbon market as well as those policies and regulations may not be the main reasons for the improvement in the market's liquidity. Therefore,

Table 12 Carbon finance methods: Hubei versus Guangdong Province

Guangdong Province		Hubei Province	
Product	Practical case	Product	Practical case
Carbon Emission Allowances Mortgage Financing	On 2014 December 25, Guangzhou University City Huadian new energy company mortgage its own 15 million tons carbon emission allowance to the Shanghai Pudong Development Bank Guangzhou Branch and acquired also an of 5 million yuan	Carbon Emission Right Pledge Loan	1. On September 19th, 2014, Hubei Yihua Group got loans from Industrial Bank of 49 million yuan by mortgaging its 2.1 million tons emission allowance 2. On November 25th, 2014, China Construction Bank Hubei Branch signed an agreement of 0.3 billion yuan with Wuhan Huaneng Power Generation Co., Ltd. Meanwhile. Everbright Bank Wuhan Branch signed an agreement of 0.1 billion yuan with Hubei JinAo Chemical Technology Co., Ltd. 3. On August 25th, 2015. China Import and Export Bank Hubei branch signed an agreement of 0.1 billion yuan with Hubei Yihua Group
Carbon Trading Coloration overdraft	On 2014 December 25, Guangzhou University City Huadian new energy company acquired a carbon trading corporation overdraft of 5 million yuan from Shanghai Pudong Development Bank Guangzhou Branch	Carbon Fund	Hubei public the China first carbon market fund on November 26th 2014. Huaneng Group and Lion Fund management Co., Ltd jointly issued the scale of 30 million "carbon emissions permits special asset management plan" fund aiming to invest in allowance trading

(continued)

Table 12 (continued)

Guangdong Province		Hubei Province	
Product	Practical case	Product	Practical case
Carbon Allowance Repurchase	\	Carbon Crowd-funding	On July 24th, 2015, CHEEX raised 0.2 million yuan in 5 min, which was use for CCER development of rural biogas digesters. This project helped the farmers increase income by revitalizing their carbon asset and benefit the investors with CCER emission reduction or certificate of honor
EA-SCP	\	Carbon Asset Custody	1. On December 8th, 2014. Hubei Xingfa Chemical industry Co., Ltd successfully entrusted its 1 million's tons carbon allowance 2. On December 22nd, 2014. Hubei Yihua Group signed agreement to entrust its 1.008 million tons carbon allowance
Low-carbon Development Fund	On October 26, 2015, Guangdong provincial development and Reform Commission and provincial finance department arranged 100 million yuan's low-carbon development fund, and commissioned Guangdong Yueke Financial Group Co., ltd. to manage	Carbon Financial Credit	By the end of November, 2014. Hubei has signed credit agreement of the "low carbon industry development and Hubei carbon financial center construction" with China Construction Bank, Minsheng Bank, Shanghai Pudong Development Bant and the Industrial Bank, the total size has reached 800 billion yuan

(continued)

Table 12 (continued)

Guangdong Province		Hubei Province	
Product	Practical case	Product	Practical case
		Carbon Bond	On November 26th, 2014, Minsheng Bank Wuhan Branch signed the domestic largest carbon bonds intentionality cooperation agreement with Hubei Huadian Power Generation Co,. Ltd for 2 billion yuan

Source http://www.cnemission.cn

policies and regulations cannot effectively stimulate enterprises to participate in market trading and are not entirely conducive to enhancing the liquidity of Guangdong's carbon market.

Second, methods of calculating and issuing carbon allowances are some of the main factors influencing the liquidity of the Guangdong carbon market. Guangdong has made several adjustments to its carbon allowance scheme, such as increasing the quantity of emission-controlled enterprises, expanding the scope of industries that suit the baseline method, reducing the quantity of paid allowance, cancelling the base bidding price and linking the price of the primary market and the secondary market, and abandoning the limitation that enterprises could only buy allowances from the primary market. These many measures together made the participants in Guangdong's carbon market more diverse and the relationship of supply and demand more balanced. Consequently, the liquidity of Guangdong's carbon market has improved. However, compared with Hubei ETS, the total allowances of Guangdong are still relatively loose and the enterprises rather concentrated.

Third, the microstructure of Guangdong's carbon market is another factor greatly influencing its liquidity. The implementation of linking the prices of the primary and secondary markets in Guangdong was helpful in reducing the fragmentation of the two markets, which as a result improved the expectations of enterprises and institutional investors of the secondary market and their participation in it. After CEEX launched a series of preferential measures on transactions, the attractiveness of the secondary market to investors was further enhanced. Additionally, comparing the trading rules of Guangdong and Hubei, Hubei's barriers to market entry are quite low, and Hubei has crafted extremely attractive trading incentives. Creating a huge scale of individual investors and institutional investors may be the main reason why Hubei's market activity exceeded Guangzhou's. Therefore, although the factors influencing investors to participate in the Guangdong carbon market vary, in terms of expanding the number of participants, the effectiveness of Guangdong's relevant measures is still not obvious.

Fourth, financial innovation that Guangdong undertook in compliance year 2015 was not the main factor influencing improvements in its carbon market. From a historical perspective, Guangdong is speeding up its carbon financial innovation, launching two carbon financial products at the same time and greatly making up for its previous lagging behind on carbon finance. However, whether or not the two products were launched and how the market liquidity performed have no obvious positive correlation characteristics. Comparing the carbon finance of Guangdong and Hubei, there is no doubt that the development of carbon finance has a great influence on the market liquidity. Hubei's carbon financial products are diverse and involved in low-carbon financing, carbon asset management, market trading and the other aspects. Each product has been operated successfully on a considerable scale, and many carbon financial products were developed hand in hand. Consequently, a virtuous cycle of carbon market and carbon finance has been created. This may be the main reason for the large difference in liquidity between Guangdong and Hubei. Therefore, Guangdong's carbon financial structure must be optimized and scaled up.

References

1. Barragán-Beaud Camila, Pizarro-Alonso Amalia, Xylia Maria, Syri Sanna, Silveira Semida (2018) Carbon tax or emissions trading? An analysis of economic and political feasibility of policy mechanisms for greenhouse gas emissions reduction in the Mexican power sector. Energy Policy 11(122):287–299
2. Duro Juan Antonio, Teixidó-Figueras Jordi (2014) World polarization in carbon emissions, potential conflict and groups: an updated revision. Energy Policy 11(74):425–432
3. Allcott H, Mullainathan S (2010) Behavior and energy policy. Science 327(5970):1204–1205
4. O'Mahony Tadhg (2013) Decomposition of Ireland's carbon emissions from 1990 to 2010: an extended Kaya identity. Energy Policy 8(59):573–581
5. Burtraw Dallas, Kahn Danny, Palmer Karen (2006) CO_2 allowance allocation in the regional greenhouse gas initiative and the effect on electricity investors. Electricity J 19(2):79–90
6. Rahmana Mohammad Mafizur, AbulKashem Mohammad (2017) Carbon emissions, energy consumption and industrial growth in Bangladesh: empirical evidence from ARDL co-integration and Granger causality analysis. Energy Policy 110(11):600–608
7. Gagelmann F (2007) The influence of the allocation method on market liquidity, volatility and firms' investment decisions. Geol Bull China 69–88
8. Daskalakis G (2018) Temporal restrictions on emissions trading and the implications for the carbon futures market: lessons from the EU emissions trading scheme. Energy Policy 8(115):88–91
9. Black Fisher (1971) Toward a fully automated exchange. Financ Anal J 27:10–21
10. Kyle AS (1984) A theory of futures market manipulation. The industrial organization of futures markets

11. Martin P (1975) Analysis of impact of competitive rates on the liquidity of NYSE stocks. Economics Staff Paper, Securities and Exchange Commission
12. Hui B, Heubel B (1984) Comparative liquidity advantages among major U.S stock market. DRI Financial Information Group study series

Towards Sustainable Energy

Transitions of Coal-Fired Power Plants in China: A Case Study in the Greater Bay Area

Hao Tan and Jingyan Fu

Abstract The transition into renewable and sustainable energy of China requires retirement of the legacy energy infrastructure and the development of new ones for sustainability. In particular, China being similar to other major economies of the world needs to gradually retire its coal-fired power generating facilities in a timely manner. Through a case study, this chapter looks into the distinct challenges and opportunities in the midst of transition of the coal-fired power sector into more clean energy facilities in China within the context of the Greater Bay Area. Implications for the future developments in the sector are discussed.

Keywords Coal-fired power · Fossil fuel · Transition · Greater Bay Area · Legacy

1 Introduction

Climate change is likely to be one of the greatest threats to global economic security and social stability in the course of the twenty-first century. The global economy's willingness and ability to reduce carbon dioxide emissions and control greenhouse gas concentrations will be crucial for climate stabilisation. Many suggest that a transition to a low-carbon economy would be an important step towards meeting this demand for climate stability [1, 2].

The efficiency of conventional fossil power plants is a strong function of the steam temperature and pressure. Research to increase both has been pursued worldwide, since the energy crisis in the 1970s. The need to reduce CO2 emissions has recently provided an additional incentive to increase efficiency [3].

A transition of the fossil-fuel-based economy is underway. The choice of fuels and technologies has ramifications far beyond the profit and loss statements of the people deciding them. In that case, decentralised decision-making will not reach a desirable

H. Tan
Newcastle Business School, University of Newcastle, Newcastle, Australia

J. Fu (✉)
College of Economics, Jinan University, Institute of Resource, Environment and Sustainable Development, Guangzhou, China
e-mail: tfjyan@jnu.edu.cn

J. Fu and A. W. Ng (eds.), *Sustainable Energy and Green Finance for a Low-carbon Economy*, https://doi.org/10.1007/978-3-030-35411-4_5

overall result. Planning and coordination are essential to tackle these problems [4]. To meet the goal established in the Paris Agreement, namely to keep a global temperature rise this century well below two degrees Celsius above pre-industrial levels and to pursue efforts to limit the temperature increase even further to 1.5 °C, the percentage of coal-based electricity in the energy system, both nationally and globally, needs to be substantially reduced.

While many current efforts are focused on stimulating new green options, weakening the cultural, political, economic and technological dimensions of fossil-fuel-related industries is just as important as stimulating green options. The bi-directional causality suggests that both processes are two sides of the same coin [5].

In January 2019, Germany's Commission on Growth, Structural Change and Employment (or the "Coal Commission" as it is commonly known) published a report proposing that the country would stop burning coal for electricity by no later than 2038. As a major energy policy initiative in the largest economy in Europe, Germany's phase-out of coal has attracted widespread attention in the world. The announcement of Germany is significant because coal-fired power currently still accounts for 35% of the country's total electricity generation—a substantial share which is only matched by that based on renewable sources such as wind and solar. Meanwhile, Germany is a country with abundant coal reserves, which has played an important role for building the country's formidable manufacturing industries during the past 200 years. The decision of Germans to move away from relatively cheap coal power is a significant development in the country's energy transition.

Germany is not the first country to declare the phase-out of coal. In 2017, at the United Nations Climate Change Conference ("COP23") held in Bonn, the UK and Canada initiated a Powering Past Coal Alliance, which includes 32 national governments (such as France, Denmark, Italy, Mexico), 22 sub-national governments (such as Australian Capital Territory, State of California) and 28 businesses or organisations as of March 2019. These countries, regions and organisations have all planned to eliminate coal-fired power generation within the next few years (Table 1). It is further estimated that the annual decommissioned coal-fired power generation capacity will overpass the annual new installed coal-fired power generation capacity globally by 2022, after which the total coal-fired power generation capacity worldwide will decrease.

Although requiring further research, it appears that in many cases the process of the energy transition was not smooth. Transitions have often depended on the timing and influence of broader external landscape forces. Both the price of the energy service and the price of the energy matter [6]. Replacing the current global energy system relying overwhelmingly on fossil fuels by biofuels and by electricity generated intermittently from renewable sources will be necessarily a prolonged process [7].

In China, a transition of the coal-fired electric power sector is driven by multiple factors. First, China has made a commitment in its pledge to the Paris Agreement to peak its CO2 emissions by 2030, and to increase the share of non-fossil energy in the total primary energy supply to around 20% by that time. In order to meet those international obligations, a significant reduction of the consumption of fossil

Table 1 Selected countries that are to phase out coal-based power generation as of April 2018

Country	Region	Installed capacity of coal power (MW)	Ratio coal-based power in electricity supply (%)	Year of total phase-out of coal power
Austria	Europe	635	5.8	2025
UK	Europe	14,442	7.0	2025
Denmark	Europe	2805	19.6	2030
Ireland	Europe	915	11.9	2025
Italy	Europe	9180	11.8	2025
Canada	North America	9458	9.0	2030
Netherland	Europe	4837	29.2	2029
New Zealand	Asia-Pacific	500	2.8	2022
Portugal	Europe	1978	25.5	2030
Finland	Europe	1836	9.9	2029
France	Europe	3526	2.5	2022
Sweden	Europe	252	1.2	2022

Source Energy Research Club based on data from Powering Past Coal Alliance and Europe Beyond Coal

fuels, especially coal, is needed in the country. Second, as the single-largest source of air pollutants in China, coal has caused considerable local environmental concerns. In 2013, in face of mounting concerns over severe smog problems, the Chinese government released its Action Plan for Air Pollution Prevention and Control, with ten major measures to tackle the issue.[1] A number of specific policies have been introduced which were directly targeting the use of coal at the national level. For example, the share of coal consumption in the total energy consumption was to be reduced to 65% in 2017 from a level of about 68% in 2013.

Meanwhile, the Action Plan also required that the coal consumption was to be reduced in absolute term in several key regions in China, notably the three biggest city clusters, including the Beijing–Tianjin–Hebei area, the Pearl River delta and the Yangtze River delta. The Action Plan seems to be effective. According to the latest energy statistics from BP (2018), the share of coal consumption in China accounted for just over 60% of the total energy consumption in 2017. In those three key regions, measures in the Action Plan have also been implemented successfully during the four years between 2013 and 2017. In 2018, a new Action Plan was issued, which follows the same principle of reducing coal use in China's energy system.[2] Third, China has

[1] See http://www.gov.cn/zwgk/2013-09/12/content_2486773.htm.

[2] See http://www.gov.cn/zhengce/content/2018-07/03/content_5303158.htm.

also faced considerable excess industrial capacity during recent years, especially in the areas of steel, coal mining and coal-fuelled power generation.[3]

According to a study by researchers at the State Grid Co. Energy Research Institute, the country would only require 980 GW of coal power capacity at maximum by 2020.[4] Even in an event that no approvals would be granted for any coal power projects from now on, the country would have an excess of coal power generation capacity at a level between 220 and 420 GW by 2020 if all approved projects are to be completed and put into production. Another more conservative estimate by researchers at the North China Electric Power University predicts that the excess of coal power capacity will reach 140 GW by 2020 [8]. Taking 300 MW as the average capacity of one coal power generation unit, and a typical coal power plant running two such units, the country needs to shut down a total of over 230 medium- and large-sized coal power plants, which would involve about RMB 2.45 trillion (or about US$365 billion) of assets.

To address the issue, the Chinese government has introduced a number of policies, such as the "Notice on orderly development of coal power" in 2016, and notices on "Early warning of coal power planning and construction" in several consecutive years since 2016.[5]

There are significant economic, social and environmental implications of the transitions in the coal power sector in China. This chapter will examine some of the distinct challenges and opportunities facing the transition of the coal power sector in China, in particular in the context of the Greater Bay Area. It discusses the experiences of coal power closures in two companies located in the area based on our field study as well as the desk research. The case study reveals that their experiences have been shaped by not only the factors at the regional and national levels, but also those at the firm level such as their respective ownership structures and business portfolios.

2 Specific Challenges and Opportunities in Coal Power Exit in China

The transition of fossil power plants in China imposes many challenges that are common across countries. Those challenges are for individual firms of concern, and other key stakeholders involved, including local governments, the communities and employees in those firms. However, the distinctive institutional and economic characteristics in the Chinese power sector also result in specific challenges and opportunities in its transition. There seem to be at least three distinct challenges

[3]See, e.g., a news report published in the website of China's central government at http://english.gov.cn/premier/news/2017/05/10/content_281475651310534.htm.

[4]http://www.cnenergy.org/sylb/tj/201704/t20170414_443421.html.

[5]See http://www.ndrc.gov.cn/zcfb/zcfbtz/201708/t20170814_857567.html; http://zfxxgk.nea.gov.cn/auto84/201805/t20180524_3186.htm.

facing exit of coal-fired power stations in China, in particular in the Greater Bay Area.

First, the scale of excessive coal power capacity that China is substantially larger than any other country in the world. By the end of 2018, the coal-fired power generation capacity in China totalled 1010 GW, which supplied about 65% of the total electricity in China. The NDRC aimed to remove 20 GW of technologically outdated capacity by 2020, which is equivalent to half of the current total coal power capacity in Germany. According to a recent report by CoalSwarm [9], China currently has a total of 260 GW of coal-fired power projects under construction, almost as much as the total coal-fired power generation capacity in the USA. Many of those projects were initiated after 2014, when the approvals were decentralised from the central government to governments at the provincial level. These capacities to be added on top of China's existing coal capacity would push the country's total coal power generation capacity to a level of 1270 or 170 GW larger than what Chinese government aims to cap by 2020–1100 GW of coal-fired power capacity—suggesting the country is having a severe overcapacity problem in the sector.

Meanwhile, the environmental problems associated with coal power generation have brought greater pressures on local governments, and consequently their strong intention to shut down more coal power plants in their areas. For example, in the two policies that almost every local government is required to develop at the provincial or even city levels, including the "Action Plan to Win the War to Defend Blue Sky" and the "Action Plan for Comprehensive Management on Air Pollution Prevention and Remedy", explicit targets are often highlighted in regards to closures of local coal power plants. The environmental pressure and the overcapacity problem make exit of coal-fired power stations in China particularly urgent.

Second, there seems to be largely a lack of coordinated approach and policy support for the exit of coal power capacity. According to our field study, the closures of coal power plants have been mainly driven by administrative orders and usually upon short notice. Due to a usually short time frame between the decision and the execution, the communications among the government, enterprises, workers and other stakeholders were often far from adequate. Unlike the steel and coal mining industries where companies receive financial support from a fund specifically set up by the central government, no such support has been provided to coal power companies for their capacity reduction activities. The costs in association with closures of plants are usually borne by the companies themselves. Further, new environmental and social problems might arise due to the closure of coal-fired power stations in certain areas. For example, the closures of coal power plants in advanced regions in China have not always translated to the greening of the energy system in the country, because those areas may import more electricity from other regions where electricity generation can be coal-based. For example, the western regions in China have seen a rise of construction activities of coal power plants in 2018.

By contrast, the "Coal Commission" in Germany was established by the Federal government in June 2018, with commission members representing a range of stakeholders including the industry, regions, environmental organisations and academia. The lengthy negotiations led to an agreement that was passed with a vote of 27:1, and the outcome seems to be widely welcome by various parties, such as environmental organisations, coal-producing regions, coal power plant workers and the user groups of electricity.

Third, most power generation companies are not willing to close their coal power plants voluntarily. Many of them have shown strong resistance even under pressures of forced closures, with a few exceptions. A comprehensive and strategic plan on the transition does not seem to be on the agendas within many of those firms either. Strategic management scholars such as Michael Porter have outlined three types of "exit barriers", including the economic, strategy and managerial barriers [10]. Economic barriers are mainly arisen from durable and specific assets that can hardly be repurposed, as well as the costs associated with layoff of workers. The strategy barriers are caused by the interrelatedness of different businesses within the corporate, in that the exit of one business may have negative impacts on the business in other areas. Finally, the managerial barriers are mainly due to a lack of information for making proper decisions, and conflicting goals between the firm and the managers, who would be unwilling to scale down the business out of the consideration of their own interests.

In 2018, nearly 50% of China's thermal power companies suffered a loss. However, among the companies we visited, few have been willing to decommission their plants voluntarily, even when the power generation units had reached or were approaching their retirement ages. The three types of exist barriers are all reflected in those cases. For example, during recent years, power generation companies have made substantial investments in desulphurisation and denitrification equipment in order to meet the government's new requirements in this area. These investments have become sunk costs. Although the coal power business is hardly profitable, the power generation firms would suffer a greater loss if the units are to be decommissioned. In addition, since electric power generation business requires approvals by various government departments at different levels, an access to this specific business has always been regarded as a valuable resource which the companies are reluctant to give up easily. The costs in relation to layoff of workers and managers are also an important consideration, which creates a major challenge for firms to move ahead with decommission if proper financial support from external sources is not in place.

Despite those challenges, there seem to exist several important opportunities of which China, and especially its advanced regions such as the Greater Bay Area, can potentially take advantage to make a transition in the coal power sector effective.

First, although the scale of coal power capacity in China that needs to be reduced is stunning, the objective is feasible technically and affordable economically. The country has the world's largest electricity generation capacity, at a total of 1900 GW. The electricity capacity and generation based on renewable energy sources has grown rapidly, in particular solar and wind. The country has also developed advanced grid networks; and the development of leading technologies such as ultra-high-voltage ("UHV") electricity transmission has made long-distance transmission more efficient. Meanwhile, the structural changes in the Chinese economy during recent years have resulted in a fall in energy-intensive manufacturing activities in China; and the capacity to supply electricity is currently far greater than the demand. For example,

China's thermal power plants worked only 4361 h in average in 2018, less than 50% of the full utilisation. Based on those factors, it seems that an orderly exit of coal power at a relatively large scale is unlikely to cause risks to China's energy security.

Second, while there is currently a lack of systematic and coordinated approach in handling coal exit in China, policy coordination in this area, if designed well, is possible because coal power exit is of the interests of the key parties involved in this development, including the central government, local governments and the firms. From the perspective of the central government, coal power exit is not only important for easing overcapacity, a central economic policy during recent years, but also a critical measure to help China fulfil its international commitment to combat climate change. Especially considering this is an area where there are considerable attentions from international media and environmental groups, it would be unwise if China loses its ground on this issue. From the perspective of local governments, closures of coal power plants could bring environmental benefits, in particular improved air quality which is one of the major concerns of local people these days. Meanwhile, land used by the power plants could be repurposed for other development, which would help upgrade the image of the city. For power generation companies, although closure of the plants is usually a challenging process, many of the companies are suffering economic losses from coal power business. An early closure of their plants would be beneficial for many of those companies as a transition strategy to negotiate for better terms with governments and other stakeholders.

Finally, a main obstacle facing the power generation companies in coal power exit arises from the costs in relation to resettlement of their workers. However, in many cities in the Greater Bay Area, there is usually considerable value attached with the land currently occupied by the coal power plants and the related transmission facilities. In fact, to access to those pieces of land has often been an important motivation for the local governments to push for the closure of coal power stations in their areas. Should proper arrangements be made, the value derived from repurposing the land would be sufficient to cover the related costs in the closure of those plants.

3 Case Study: Two Tales of Coal Power Exit

3.1 Background

The coal power exit in China has been driven by a number of economic, policy and environment factors. In the Greater Bay Area, a number of coal power plants have been in particular affected by the recent plan to develop the region as a new economic engine as well as an exemplar in the country's effort to promote ecological civilisation. According to the Outline of Development Plan for Guangdong-Hong Kong-Macao Greater Bay Area (thereafter "the Outlines"), to achieve those goals, the structure of energy supply in the region needs to substantially change.[6] Specifically, a clean, low-carbon, safe and efficient energy supply for the region as suggested in the Outline requires an increase in the use of renewable energy and natural gas, and meanwhile a significant reduction in coal consumption. Such a strategy implies an aggressive coal power exit from the region.

In Guangdong province, additional measures have already been implemented under its "Action Plan to Win the War to Defend Blue Sky 2018–2020" to reduce the use of coal.[7] For example, the Action Plan requires a cap of the percentage of coal in the primary energy consumption in the province at 37%. According to the Action Plan, the total consumption of coal needs to be below 70 million tons in the Pearl River delta area, or 10 million tons less than the 2015 level.

In the area of coal power generation, the Action Plan explicitly requires the following: the province is to strictly limit the production capacity in pollution-intensive and emission-intensive industries; to firmly close coal-fired power units that are at a capacity under 300 MW and do not meet the environment, energy and/or safety standards; to press earlier retirements of long-serving coal-fired generating units.

The Guangdong Provincial Development and Reform Commission needs to coordinate with the State-owned Assets Supervision and Administration Commission of Guangdong Province and the Guangdong Energy Group Co., Ltd, to decommission a number of coal-fired power generating units at a total capacity of 3240 MW by the end of 2020, including those in Guangzhou Power Plant, Guangzhou Wanglong Thermal Power, Shajiao Plant A and Shajiao Plant B.

However, there are challenges to achieve those goals. Fossil fuels currently account for about 70% of the current energy mix of Guangdong province. Meanwhile, the energy resource endowment in the region is limited, resulting in a high level of energy dependency on inter-provincial or even inter-country imports. For example, about one-third of electricity demand in Guangdong Province is supplied through the West-East Electricity Transmission Project, where the electricity is produced in western provinces such as Guizhou and Yunnan, and transmitted to Guangdong for end users. Therefore, the transition of the energy system in the region, and in particular the closures of coal power plants, faces challenges at multiple dimensions. The "energy trilemma" in the decision-making and implementation process of coal power exit in this context involves such considerations as energy/power supply security and costs, environmental concerns of different stakeholders at the local, national and international levels, and local and national economic growth.

The two coal power plants involved in our case study were both located in one of the cities in the Greater Bay Area but owned by different electricity companies. In 2018, both coal power plants were required by the provincial government to be closed by 2019.

As part of the Greater Bay Area strategy at the local level, the local city government plans to establish new economic engines to replace the "old", energy- and pollution-intensive industries. In particular, a new development area/city has been planned in

[6]http://www.gov.cn/zhengce/2019-02/18/content_5366593.htm (in Chinese).

[7]http://www.gd.gov.cn/zwgk/wjk/qbwj/yf/content/post_1055797.html (in Chinese).

the city, which is positioned as a cooperation platform of the Greater Bay Area, a leading area for Guangdong–Hong Kong–Macao Collaborative Development, and a core innovation platform for Guangzhou–Shenzhen–Hong Kong–Macao Science and Technology Innovation Corridor. The two coal power plants were both located in this new area that is currently under plan and development.

From the local government's point of view, such a strategy can hardly move forward without relocating those coal power plants, as their existence becomes an obstacle for the development of key transportation corridors and other major infrastructure and industrial projects. In addition, the local government anticipates an introduction of advanced manufacturing industries into this area, where the tax yield per unit of land is anticipated by the local government to be much higher than the tax yield at the current level from the coal power generation businesses. The local government has also a strong interest to develop a number of public facilities such as parks and museums at the location in order to upgrade the image of the city.

3.2 The Two Main Companies

Company A in our case study is a large state-owned enterprise and a key electric power supplier in the province. Although the company has undergone several reforms and restructuring in the past, its governance is currently still featured with a highly concentrated ownership structure, with 76% of the shares owned by the Provincial State-owned Assets Supervision and Administration Commission, and the remaining 24% owned by one of the major national utility companies.

The company is almost solely reliant on electricity generation and sales, which accounts for over 99% of its revenue. While the company has entered in other areas of energy business during recent years, such as power generation based on renewable sources including wind, solar and biomass, coal-based power generation currently still accounts for 83% of total electricity generation of the company. Further, as a major state-owned utility company in the province, the security and costs of the electricity supply for the province is one of the main responsibilities of the company. A high level of assets and the large number of employees also brought challenges to the company during its coal power plant closure process.

Company B is also a state-owned enterprise. However, the company's ownership is more diffused. The company was publicly listed in 2007. While the State Assets Supervision and Administration Commission of a local city is the controlling shareholder of the company, it owns less than 50% of the company shares, and the rest is owned by other shareholders including a national energy company. The company has also introduced an employee stock ownership plan.

The business of Company B is also more diversified. While electricity generation and sales are the main business of the company, accounting for 82% of the total revenue in the last year, the business portfolio of the company involves such businesses as gas sales, waste processing and real estate development. Within the electricity generation business, the coal-based power generation accounts less than

50% of Company B's total electricity generation, and the rest being contributed by gas (25%), hydro-power (8%), solar (9%), wind (6%) and waste incineration (2%).

The coal power plants of Company A and Company B are co-located at a central location of the Pearl River delta and are both parts of a power generation base which has played a significant role in powering the economic activities in one of most economically advanced regions in China. Together with another power plant, the two coal power plants act as a key support for the electricity grid in the region. The coal power plant owned by Company A (thereafter "PP-A") included five power generation units and had about 2,000 employees before the closure. The coal power plant owned by Company B (thereafter "PP-B") was originally built by a foreign company through a build–operate–transfer arrangement. For the first 14 years, PP-B was fully operated by the foreign company based on its own management practices, before the plant was transferred to the local SoE. The company had about 360 employees before the closure.

While both companies have faced considerable challenges arisen from the forced closure of their respective power plant, their responses and experiences somewhat differ. Company A seems to have largely taken a reactive approach, hoping the government at the provincial level to change or delay the plan to close PP-A. When such a hope faded, the company did not have a clear strategy, and the company has largely been viewed as being uncooperative in its subsequent dealing with the issue by the local government. By the time of our field study, the company did not seem to have actively engaged in negotiations with key stakeholders, including the local city government, for seeking a better condition in a post-closure arrangement. By contrast, while Company B was also originally against the decision to close its PP-B plant, the company had managed to negotiate with the government for an approval which allows the company to build a new natural gas-based power plant in a nearby city. Given the smaller size in its workforce, the relocation of employees and managers from PP-B also seem to be more effective.

As a major energy SoE in the province which has enjoyed being a dominant market player for years, Company A seems to have a strong risk-averse culture and considerable inertia against a change of its business portfolio. Its existing competitive advantage based on the economy of scale in coal fire power generation business and the related technological, physical and human resources in this area also creates an exit barrier. The corporate governance of Company A also does not provide motivations for its management and workers to support the transition, especially given a number of uncertainties in relation to the closures of coal power plants.

Compared with Company A, coal power exit is more compatible to Company B's corporate strategy which is aimed to become a comprehensive energy company, with a focus on developing other energy businesses beyond coal power. The employee stock ownership plan seems to have also facilitated better involvement of employees in the reorientation of strategy, and a swifter decision-making according to the change of the business environment. All those factors seem to have contributed to the different closure processes we observed in the two coal power plants.

4 Discussion and Concluding Notes

In face of external pressures arisen from energy transition, incumbent energy firms display different behaviours. Some incumbent energy firms have responded to the transition proactively, for example, by engaging in new energy technologies, while many others delay their response, and others resist transitions by, for example, forming political coalitions [11]. van Mossel et al. [11] draw from five theories in management and organisation studies to examine those responses, including Institutional Theory, the Behavioural Theory of the Firm, the Resource-Based View, Resource Dependence Theory and Organizational Ecology. Those theories have shed light (in different ways) on the two main questions in regard to the antecedents and performance implications of various incumbent behaviours, respectively. Those findings are summarised in Table 2.

In our study on incumbent coal power generation firms in the Greater Bay Area in China, we have also observed different exit behaviours by firms under the same pressures from the government to close their plants. Those behaviours seem to be subject to factors at the firm, regional and national levels. At the firm level, the corporate governance and ownership structure, and the existing business portfolios seem to have played important roles. At the regional level, the local political economy and especially a heavy use of industrial policy and strong government interventions in economic and social affairs with an aim to upgrade the economic structure in the region in competition with other regions are some of the major determinants in driving the decision and shaping the process of the coal power exit. At the national levels, the country's international commitments to climate change are also an important consideration.

Compared with the increasing research on incumbent firms in energy transitions in the developed country contexts, especially that based in cases in Europe, research on behaviours of incumbent firms in energy transitions in China is still rare. It is largely an unexplored area in the transition literature to study particularly destabilisation of incumbent firms in energy transitions at the firm level taking specific economic and institutional environments of China into the consideration. Our study, albeit in an early stage, makes a first step to fill this gap.

Practically, the transition of the Chinese coal power sector, in particular the closures of a large number of coal power plants necessary for the transition, is having substantial implications to not only the national energy system and regional economy, but also the global environment. According to the policy issued by the National Development and Reform Commission (NDRC), China needs to cancel or postpone 150 GW of coal-fired power stations that have been planned, and remove 20 GW of existing coal-fired power generation capacity during the "Thirteenth Five-Year Plan" period ending in 2020. Future international comparative studies on the barriers, motivations and strategies in the process of coal power exit will inform and facilitate a more effective and just energy transition in China.

For example, in Australia, coal power stations are not forced to exit by the government. However, several major power generation companies have taken the initiative

Table 2 Relationship between incumbent characteristics and behaviour during transitions

Theory	First to enter niches	Follow into niches	Delay the transition	Remain inert
Behavioural theory of the firm	Incumbents who do not meet their aspirations, in the absence of a niche that is perceived to be successful. Incumbents with a high amount of slack	Incumbents who do not meet their aspirations, in the presence of a niche that is perceived to be successful		Incumbents that meet their aspirations
Resource-based view	Incumbents with dynamic capabilities Incumbents with excess resources that can be exploited in a niche	Incumbents with dynamic capabilities and resources that fit better with a niche than the regime	Incumbents that can raise isolating mechanisms	Incumbents without dynamic capabilities or excess resources
Resource dependence theory	Incumbents that control resources that allow them to increase the control over their environment in a niche	Incumbents that are dependent on powerful firms that support a niche	Powerful incumbents that support the incumbent socio-technical regime	Incumbents that are dependent on powerful firms that support the incumbent socio-technical regime
Institutional theory	Incumbents that are exposed to multiple conflicting institutions Incumbents whose technically efficient practices deviate strongly from institutional demands	Incumbents whose legitimacy is higher in a niche than in the regime	Incumbents whose organisational privileges and social position are threatened	Incumbents that are only exposed to the institutions from the incumbent socio-technical regime. Incumbents whose legitimacy is higher in the incumbent socio-technical regime than in a niche
Organisational ecology	Incumbents with a history of innovative organisational change	Incumbent generalists Small incumbents		Incumbent specialists Large incumbents

Source van Mossel et al. [11]

and have planned the closure of their coal power plants voluntarily within the next few years. Such actions are mainly based on considerations that it would be economically beneficial to undertake a well-planned exit of those power generation plants than to continue to operate and maintain those facilities. This is especially the case given future climate policies and public oppositions will increasingly place uncertainties and pressures for the businesses in this area. Some of the companies have introduced the notion of "transition management" [12], which helped overcome some of the challenges in the planning process of the closure. For example, AGL, one of the largest power generation companies in the country, has announced the company will close two of its large coal-fired power stations in the Hunter region, which currently supply almost 30% of the electricity used in New South Wales, the largest state in Australia. In the process, the company has initiated a Hunter Energy Transition Alliance which involves local governments, communities and research institutes. The Alliance has been used as a platform of communication, as well as for a number of research projects concerning the preparation and implementation of the closure. Companies and other related parties in China could look into this experience for ideas and inspiration.

Acknowledgements The authors thank Guangdong Energy Saving Centre and the companies that they visited during the field study, for their help and assistance to this research.

References

1. Foxon TJ, Kohler J, Oughton C (2008) Innovation for a low carbon economy: ¨economic institutional and management approaches. Edward Elgar Publications, Cheltenham and Northampton, MA
2. Grubb M, Jamasb T, Pollitt MG (2008) Delivering a low-carbon electricity system: technologies, economics and policy. Cambridge University Press, Cambridge
3. Viswanathan R, Sarver J, Tanzosh J (2006) Boiler materials for ultra-supercritical coal power plants—steamside oxidation. J Mater Eng Perf 15(3):255–274
4. Allen RC (2012) Backward into the future: the shift to coal and implications for the next energy transition. Energy Policy 50:17–23
5. Turnheim B, Geels FW (2013) The destabilisation of existing regimes: confronting a multi-dimensional framework with a case study of the British coal industry (1913–1967). Energy Policy 42:1749–1767
6. Fouquet R, Pearson P (2012) Past and prospective energy transitions: insights from history. Energy Policy 50:1–149
7. Smil V (2016) Debating energy transitions: a dozen insights based on performance. Energy Res Soc, Sci in this issue
8. Yuan JH (2017) China's coal power during the 13 and policy. Presentation. North China University of Electricity. Available at https://www.china5e.com/download/20161101-nrdc/1101-YuanJiahai.pdf
9. Shearer C, Yu A, Nace T (2018) Tsunami warning: can China's central authorities stop a massive surge in new coal plants caused by provincial overpermitting? Report of CoalSwarm. Available at http://coalswarm.org/trackers/a-shrinking-coal-plant-pipeline-mide-2016-results-from-the-global-coal-plant-tracker/

10. Porter ME (1976) Please note location of nearest exit: exit barriers and planning. Calif Manag Rev 19:21–33
11. van Mossel A, van Rijnsoever FJ, Hekkert MP (2018) Navigators through the storm: a review of organization theories and the behavior of incumbent firms during transitions. Environ Innov Societal Transitions 26:44–63
12. Loorbach D, van der Brugge R, Taanman M (2008) Governance in the energy transition: practice of transition management in the Netherlands. Environ Technol Manage 9(2/3):294–315

Penetration of Electric Vehicles into the Greater Bay Area

Yang Andrew Wu and Songdan Nancy Guo

Abstract Green technology has increasingly penetrated our daily lives over the past decade. Electric vehicles (EVs) have been widely accepted by both corporate users and individual customers. Recently, the Chinese government has been firmly driving forward its macro-economic policy, "Made in China 2025," in manufacturing and industry. At the same time, Chinese citizens are demanding a cleaner, healthier natural environment for their urban lives. Promotion of EV has turned out to be one attractive solution for a large number of stakeholders from various sectors. This chapter begins with a review of cutting-edge EV technology development internationally. It covers smart control and driving technology, new batteries, and energy storage devices. It is followed by a comparative analysis of a range of policy incentives currently effective in the Greater Bay Area (GBA) of China. These policies from different authorities and departments need subtle cooperation in order to be fully executed. This chapter discusses multiple incentives from three aspects: technology, economics, and policy direction. The author believes that both EV manufactory and consumption in the GBA will experience steady growth in the coming years. We conclude that an upward trend of EV penetration in the GBA will positively contribute toward energy sustainability as well as the green finance of projects in the future.

Keywords Electric vehicle · Sustainable development · Energy policy · The Greater Bay Area · Green finance

1 Current Status Quo of EV Technology Development

Electric vehicles are not exactly new things. It has been more than a century since the first small-scale electric vehicle was invented. However, in the 1920s, with the discovery of enormous quantities of crude oil all over the world, gasoline prices quickly fell to affordable levels, and the construction of infrastructure such as roads and gas stations improved, making fuel vehicles more cost-effective. Since then,

Y. A. Wu (✉) · S. N. Guo
School of Professional Education & Executive Development, College of Professional & Continuing Education, The Hong Kong Polytechnic University, Kowloon, Hong Kong SAR, China
e-mail: andrew.wu@cpce-polyu.edu.hk

J. Fu and A. W. Ng (eds.), *Sustainable Energy and Green Finance for a Low-carbon Economy*, https://doi.org/10.1007/978-3-030-35411-4_6

no major breakthroughs have been made in battery and cruising technologies [1]. Consequently, electric vehicles have gradually lost their comparative advantages and have been overtaken by fuel vehicles with internal combustion engines.

After entering the second half of the twentieth century, and after two large market shocks in the oil industry, the auto market and the public began to focus on new energy vehicles. This change has been led by electric vehicles. As a result of poorer air quality since the 1990s, more and more people have examined the possibility of introducing commercial electric vehicles into our transportation system [2]. New products with lower emissions, greater efficiency, and even zero emissions have been under research and development. Under the influence of a variety of factors such as the continuous rise of global oil prices and the increasingly strong demand for environmental protection in the 2010s, the commercialization of electric vehicles has once again been put on the agenda by governments and major automobile companies.

One major difference between a fuel car with an internal combustion engine and an electric vehicle is the powertrain system. The fuel car's own powertrain system consists of the engine assembly, the gearbox assembly, and the fuel tank. The powertrain system of the electric vehicle is powered by the battery pack. The motor and the electric control system are designed so that the power battery pack replaces the fuel tank, the drive motor replaces the engine, and the electric control system replaces the gearbox [3]. There are now a large number of manufacturers of these electric vehicles worldwide. In this paper, we have selected three typical examples from China, Japan, and USA for a brief overview.

In China, BYD started business in the battery market and has a solid background in lithium iron phosphate battery technology. In 2008, BYD launched the F3DM with a lithium iron phosphate battery. The F3DM was the world's first plug-in hybrid vehicle with a top speed of 150 km/h [4]. The 100-km cruising range was also achieved in pure electric mode, and it set a world record at that time.

At the end of 2010, after nearly ten years of research and development, Nissan launched its pure electric vehicle, the Leaf, equipped with a lithium manganese battery. The first-generation Leaf had a cruising range of 160 km. Its lithium manganese battery with a capacity of 24 KWh was provided by NEC and Nissan's joint venture—AESC. The energy density reaches 140 wh/kg [5].

In the USA, Tesla launched the iconic electric racing car Roadster in 2008, and the Model S, a luxury sedan equipped with a ternary battery (nickel–cobalt–aluminum/NCA), in 2012. The Model S can reach a top speed of nearly 200 km/h. Its ternary battery is supplied by Panasonic and has an energy density of 170 Wh./kg, with a battery capacity of 85 KWh [6]. It can achieve a cruising range of nearly 500 km.

The key factor for the realization of innovative electric vehicles is the development of battery technology. It is predicted that the large-scale commercialization of all-solid-state batteries might not be realized until 2025–2030, or even later. In addition, we can be fairly sure that technological changes in high-efficiency drive systems will occur in the next five years, that is, the motor drive system will become faster, more efficient, and achieve greater miniaturization.

The history of technological progress of electric vehicles is also a history of power batteries products. By comparing the representative models of those three electric vehicles, we can see that the development and application of new battery materials, energy density, and cruising range have been improving over time throughout the industry.

The battery mainly includes five performance indicators: energy density, safety performance, cycle performance, high and low temperature performance, and cost. Among the current mainstream cathode materials, lithium iron phosphate has the best safety performance, cycle performance, and lowest cost, but has a poor low temperature performance and low energy density; lithium manganese is relatively safe in terms of safety performance, cycle performance, and cost [7]. However, poor high temperature performance and low energy density are drawbacks; ternary materials are relatively poor in safety and high in cost, but energy density is higher than the other two materials. Due to the passenger car's requirements for cruising range, the current ternary materials, especially the high-nickel ternary, are increasingly favored by the industry, and the market share continues to increase.

2 Overview of EV Promotion Policies in the UK and USA

As early as the 2000s, developed countries such as the USA and those in Europe had begun to vigorously promote the development of new energy vehicles and had launched a series of policy incentives to promote the construction, sales, and use of new energy vehicles. In fact, the concept of new energy vehicles was first proposed decades ago.

In 2009, the UK introduced the ULEV Strategy to promote the nationwide use of ultra-low emissions vehicles, especially new energy vehicles. The details of this program were completed in 2013 [8]. The significance of the plan was to lay a solid foundation for the infrastructure construction of new energy vehicles. As we know, the promotion of new energy vehicles is often subject to the progress of basic design and construction such as charging equipment. The ultra-low emission vehicle program provides policy guidance for infrastructure construction across the UK, including charging stations and peripheral support settings. In London, for example, it is expected that more than 6,000 charging piles will be completed by the end of 2019.

Further to this, the British government proposed in the air quality plan promulgated in July 2017 that it would completely end the primary market sales of traditional fuel vehicles by 2040 [9]. In October of the same year, the Ministry of Commerce and the Ministry of Energy promulgated the Clean Growth Plan [10], promising that the government will spend around a billion pounds to promote the market share and actual utilization rate of new energy vehicles. Furthermore, the British government's Climate Change Committee proposed in January 2018 that the UK completes the goal of ensuring at least 60% of new cars on the road in 2030 are ULEVs. The development of the new energy automobile industry in the UK has received unprecedented political and economic support.

In the USA, various state governments have also introduced different policy support and economic incentives, including estimates of infrastructure facilities, tax incentives, and subsidy incentives. These tend to be most prominent in coastal cities along the West Coast. According to the International Clean Energy Transportation Committee research report [11], as of the end of 2017, more than 200 cities and regions in the USA had begun to popularize new energy vehicles, including hybrid vehicles. This policy support and the economic incentives that go with it are even greater than in the UK.

More importantly, European scholars have proposed a policy system for the establishment of comprehensive new energy vehicles and related infrastructure networks. British scholar Benjamin K. Sovacool [12] pointed out that a combination of automotive technology networks and social science research, from the perspectives of technology, culture, and society, was necessary in order to accelerate the process of consumer acceptance of new energy vehicles. Swedish scholar Zeinab Rezvani and his colleagues have tried to find out why new energy vehicles have not been widely covered so far in terms of consumer behavior habits [13], consumer rational analysis, environmental attitudes, and living habits. It seems likely that consumer acceptance of new energy vehicles is a gradual and long-term process. The quarterly research report of the internationally renowned consulting firm McKinsey [14] pointed out that the penetration rate of new energy vehicles in international metropolises is gradually increasing, and that future market prospects are becoming more extensive.

3 EV Promotion Policy in the GBA of China

The industrial policy changes of China's new energy vehicles can be roughly divided into three stages, namely the government's macro-strategic planning (2001–2006), the establishment of industry access rules and comprehensive subsidies (2007–2015), and the post-subsidy stage (2016–the present).

In the first phase, the technical route of new energy vehicles has not been fully determined. The focus for a large number of countries in the world is on hydrogen fuel cells, and China is no exception. At this time, there are not many government policies. Most of the policies are macro-strategic guiding policies, focusing on the planning and development goals of new energy vehicles. For example, in 2001, China launched the "863 program of electric vehicles" and planned to build the three vertical and three horizontal new energy vehicle's development layouts [15]. In 2004, the National Development and Reform Commission issued the "Automobile Industry Development Policy," which highlighted the development of environmentally friendly technologies and electric vehicle technologies for sustainable development [16]. In 2005, the central government introduced measures to optimize the structure of the automobile industry, promote the development of clean vehicles and electric vehicles, and outline the development goals of electric vehicles over the next 25 years [17].

In the second phase, China's new energy vehicles have made major breakthroughs in key technologies. Since 2007, China's independent research and development of pure electric, plug-in hybrid and fuel cells of new energy vehicle products has progressed. At the 2008 Beijing Olympic Games, the Ministry of Science and Technology organized relevant domestic automakers to provide a concentrated display and usc of about 500 new energy vehicles for various types of energy-saving at the Olympic Games. This opened the first year of new energy vehicles in China. At this stage, the government's policies began to be refined, the industry's access conditions and corresponding norms and standards were established, and a series of financial subsidy policies were gradually introduced, laying a solid foundation for the industry to follow.

Since November 2007, the "New Energy Vehicle Production Access and Management Rules" have officially begun to be implemented in China. These rules make specific provisions for the definition of new energy vehicles, the qualifications of new energy vehicles, the conditions for production access, and the requirements for reporting. In 2009, the government issued the first subsidy policy for energy-saving and new energy vehicles, thereby subsidizing these vehicles in the public service sector. In 2010, the government subsidized the private purchase of new energy vehicles and subsidized them at the rate of 3,000 yuan per KWh. The maximum subsidy for plug-in hybrids was 50,000 RMB per EV car, and the maximum subsidy for pure electric vehicles was 60,000 RMB per car [18].

After the Chinese government began to subsidize new energy vehicles in 2009 and 2010, the domestic electric vehicle market began a more vigorous development. In 2011, domestic EV sales were only 8000 units, while sales growth in 2016 had reached 336,000 units. This accounted for nearly 45% of the global share, surpassing the USA to become the world's largest electric vehicle market [19]. 2014 is considered to be the breakout year for domestic electric vehicles. In 2015, it ushered in a full-scale upsurge in the industry. The main reason was that the development of new energy vehicles in China was lower than expected in 2010-2013, in order to complete the "Energy Conservation and New Energy Vehicle Industry Development Plan 2012." The target for the production and sales of pure electric vehicles and plug-in hybrid vehicles (including passenger cars and special vehicles) is 500,000 units by year 2020. The government introduced a number of policies in 2014 to further support the development of the industry. As a result, the industry has experienced explosive growth. The decline in subsidies has been delayed, and the list of cities to be promoted has increased, following the example of the Greater Bay Area.

In November 2016, the State Council issued the "13th Five-Year Plan" strategic emerging industry development plan. The plan has proposed that by 2020, the annual production and sales of new energy vehicles will exceed 2 million, and the cumulative production and sales will exceed 5 million. The overall technical level will keep pace with that internationally. Finally, a group of internationally competitive new energy vehicle producers and key components enterprises will be formed.

In the third phase, the state made a series of adjustments to the subsidy policy, including investigating the fraudulent behavior of some car companies, differentially adjusting the subsidy range of passenger cars and special vehicles, and introducing a long-term mechanism such as "double points" to guide transformation and upgrading of car companies.

From the results of domestic EV sales, we could observe that sales of pure electric vehicle increased from 5600 in 2011 to 257,000 in 2016, nearly 50-fold in five years. The sales of plug-in hybrid vehicles increased from 2600 in 2011 to 79,000 vehicles in 2016, an increase of around 30-fold. In general, BEV has achieved a faster growth rate than PHEV, and the proportion of BEV's total sales has remained at around 70% [20].

Regarding the passenger car, in addition to the subsidy range of the post-subsidy era, the policy has adjusted the subsidy amount for different cruising ranges. In 2016, the subsidy value of the high cruising range was greater than that in 2015. The subsidy policy aims to encourage further development of high cruising ranges and high energy density.

After the accumulation of the past three five-year plans, China's new energy vehicle industry has grown from scratch and has made significant progress in key components, e.g., vehicle integration technology, technical standards, testing technology, and demonstration operations. Powertrain control, drive motor, and power battery are the three key components for strategic research and development.

At present, the policies and regulations on new energy vehicles in mainland China are mainly issued jointly by the National Development and Reform Commission, the Ministry of Finance, the Ministry of Science and Technology, the Ministry of Industry and Information Technology, and the Energy Bureau. The policy and planning statistics that have major direct impacts on the development of new energy vehicles in the Greater Bay Area of Guangdong, Hong Kong and Macau are shown in Table 1. The table also contains comparative analysis of the key parts of each policy. The analysis focuses on the actual impact and application of the Greater Bay Area in Guangdong, Hong Kong, and Macau.

The governing bodies of the Greater Bay Area have always attached great importance to the construction of an ecological civilization and the scientific path to a low-carbon and sustainable development. "It is necessary to have Jinshan Yinshan and green water and green mountains." In fact, Guangdong Province, especially the Greater Bay Area has been at the forefront of the development of new energy vehicles. A number of important scientific and technological breakthroughs began life here.

4 The Increasing Trend of EV Penetration in the GBA

The Provincial Development and Reform Commission of Guangdong Province has pointed out that Guangdong will fully implement the national development strategy for the promotion and application of new energy vehicles (GDRC 2010). It is effectively alleviating energy and environmental pressures, and the relevant departments

Table 1 Brief summary of key policies for EV promotion in 2017 and 2018

Category	Policy	Authorities	Summary
Guideline	"Medium and Long-term Development Plan for Automobile Industry" [24]	Ministry of Industry and Information Technology (MIIT), Development and Reform Commission (DRC), Ministry of Science and Technology (MST)	By 2020, to build several domestic automobile companies into the world's leading new energy auto companies, and to increase their global market share. By 2025, to build a top-tier smart network for new energy vehicles
	"National Guideline for Vehicle Network Construction and Industrial Standard System" [23]	MIIT	Actively build a new generation of intelligent networked vehicle standard systems, to support functions such as assisted driving and automatic driving; independent research and development of safety standards, and communication standards
Tax	"PR China Vehicle Purchase Tax Law" [27]	Ministry of Finance (MoF), Taxation Bureau (TxB)	For eligible new energy vehicles, the vehicle purchase tax will be exempted or reduced.
Subsidy	"Notice on Liquidation of New Energy Vehicle Subsidy Funds in 2016" [21]	MoF, MIT, DRC	New energy vehicles purchased by non-individual users may also apply for subsidies, which require a cumulative travel of over 30,000 km. Subsidy standards and technical requirements are implemented on an annual basis in accordance with the driving permit
Facilities	"Guidance on Promoting Energy Storage Technology and Industrial Development" [26]	MoF, Energy Bureau (EnB), MIIT, DRC, MST	Multi-angle to promote the construction and development of supporting infrastructure for smart charging, power batteries, and communication base stations of new energy vehicles. Improve battery life cycle management, give power battery gradient use, and improve energy efficiency
Standards	"Promote the development plan of the automotive power battery industry" [22]	MoF, MIIT, DRC, MST	Improve the quality of new energy vehicle batteries, research and promote the application of a new generation of lithium-ion automotive power batteries; develop and test new battery systems with chemical principles

(continued)

Table 1 (continued)

Category	Policy	Authorities	Summary
	"New energy vehicle manufacturing enterprises and product access management regulations" [25]	MIIT	Increased access thresholds for new energy vehicles in research and development, design, production, sales, after-sales, safety, etc., and strengthened regulatory requirements

of the province and relevant enterprises are closely cooperating to actively promote the development of the new energy automobile industry. Supporting facilities for electric vehicles have also been vigorously developed, and the overall construction is in good condition.

According to China Electric Vehicle Charging Infrastructure Promotion Alliance statistics, as of the end of 2017, the number of new energy vehicles in the country exceeded 1.7 million, and Guangdong Province had produced more than 200,000 new energy vehicles, accounting for about one-eighth of the country. There are now 214,000 public charging piles in the country. There are about 690 charging stations in the province, 38 intercity charging stations, and more than 60,000 public charging piles. The total number of charging piles is over 80,000, and the ratio of pile to truck is as high as 1:2.4, far exceeding the national average. New energy vehicles are mainly concentrated in the Pearl River Delta region. As a consequence, the charging facilities in the Pearl River Delta are being rapidly developed, accounting for more than 90% of the province's charging facilities [28]. This basically meets the needs of charging and replenishing electric vehicles in Guangdong Province.

According to the statistics of the Hong Kong Trade Development Council, the GBA has a land area of about 55,904 km^2 and a population of nearly 70 million [29]. The regional GDP is about 1513.4 billion US dollars, and the per capita GDP is more than 20,000 US dollars. If we compare it with the New York metropolitan area and the Tokyo bay area of Japan, we see that the land area, population share, total GDP, and even per capita GDP of GBA are higher than the New York metropolitan area and the Tokyo bay area. The economic development and future potential of GBA is tremendously encouraging.

Trend 1: Public vehicle first

Zhuhai government has established a pioneer project at one bus charging station on Haihong road. The bus charging station was put into operation in October 2016. According to data obtained, in 2017 the electric vehicle charging capacity of Zhuhai Power Supply Bureau accounted for 70.3% of the electric vehicle charging capacity within the Guangdong power grid, of which electric bus charging accounted for the majority [30].

At present, all parts of Guangdong are accelerating the promotion of electric buses. "Implementation Opinions of the General Office of the People's Government of Guangdong Province on Accelerating the Promotion and Application of New Energy Vehicles" has pointed out that the proportion of renovated or newly added pure electric buses in the Pearl River Delta region shall not be less than 90% by 2020. It is required to effectively realize the large-scale and commercial operation of pure electric buses.

In addition to buses, Shenzhen is also targeting other vehicles such as taxis. The charging frequency of private cars is not high. The main application scenarios such as communities and units are still small. However, electric taxis can drive the cultivation of the public charging market more than private electric cars. Currently, Shenzhen has promoted 16,359 new energy buses, which are 100% purely electrified, except for emergency transportation capacity. The number of pure electric taxis now exceeds 13,000, and the pure electrification rate is 65%. A total of about 40,000 charging piles have been built [31].

Trend 2: Striving for innovations

Guangdong government has published the "Electric Vehicle Charging Infrastructure Planning (2016–2020)" which states that by 2020, Guangdong Province will build 1490 centralized charging stations and build about 350,000 decentralized charging piles, totaling about 410,000 in the province. A large-scale charging network for electric vehicles will be realized, and the electric vehicle will be able to travel smoothly through the Pearl River Delta.

Shenzhen has built China's first pilot project of "Smart Parking + Charging Integration" roadside charging piles. This "parking + charging" mode not only solves the car owner's cruising range anxiety, but also helps to balance the grid load and avoid the impact of large-scale centralized charging on the power grid.

The Shenzhen government has transformed one power substation at Lianhua Mountain into a multi-functional green "change + charge" power station, which covers an area of more than 3000 square meters. In addition to the original substation, the roof has also been transformed into a photovoltaic power generator. This can satisfy the fast-charging demand of 720 electric vehicles in one day.

Despite the development of charging facilities in recent years, the land issue is still the most difficult issue restricting the development of charging facilities. In particular, the legal use of land is still facing significant problems, and illegal construction is not uncommon. For example, in terms of bus charging facilities, there is no independent planning land for bus charging stations, most of which are built on non-state-owned temporary land. The huge gap in bus stations directly restricts the construction of new energy bus charging facilities. This makes it difficult for charging stations to meet daily charging and maintenance needs.

In terms of social charging facilities, the existing building parking lots and public parking spaces are limited, and there is a contradiction between the construction of charging facilities and the use of existing parking spaces. The operating income of charging facilities is problematic to achieve, which makes it difficult to fully cover the construction of charging facilities in the short term.

Trend 3: A long road ahead

New energy vehicles have not yet been widely accepted by individual consumers. An effective commercial charging facilities' operation is still being developed. The utilization rates and return rates of most public charging facilities are low, which affects the enthusiasm of operators for investing in the required infrastructure.

In order to standardize the planning, construction, and operational management of charging facilities and to promote the use of electric vehicles, the province has formulated detailed measurements in accordance with the principles of "appropriate advancement, reasonable layout and regional differences." Guangdong Province Expressway Charging Infrastructure Planning and Construction Plan (2018–2020) requires that by 2020, the province's trunk highways (all highways except the city ring road and expressway) will be "fully charged." New expressway service areas should be equipped with fast-charging piles, and reserved charging facilities' interfaces should be present at a ratio of not less than 50% of the total number of parking spaces. Relevant standards are included in highway design and acceptance specifications.

In 2017, Guangdong Power Grid established the Guangdong Electric Vehicle Charging Infrastructure Promotion Alliance to expand more than 50 influential enterprises in the industry, promote industry standardization, and build a provincial-level electric vehicle charging service via an "easy-to-charge" app on mobile phones.

The Provincial Development and Reform Commission has also made it clear that the role of the provincial charging facility alliance should be fully utilized. Expanding the influence of the charging facility alliance, and building a public information intelligence service platform, should further realize interconnection and interaction, and improve the level of charging services.

It has been found that the proportion of plug-in hybrid vehicles in the entire new energy vehicle market is increasing. In fact, compared with pure electric vehicles, plug-in hybrid vehicles can better balance mileage anxiety with the cost of pure electric vehicles and can improve the efficiency of traditional internal combustion engines. The energy efficiency of the entire vehicle is improved, and the battery usage reduced. This further reduces the vehicle weight and vehicle costs.

In the GBA, each city has its own strengths. The economic advantages have complemented each other and created huge business opportunities. For example, Guangzhou, Foshan, and Dongguan have always been important manufacturing bases, while Shenzhen has had the advantages of high-end manufacturing, innovative technology, and information technology industries; Hong Kong is world-renowned for global finance, international shipping, and offshore asset management; Macau is a popular tourism and leisure center. The complementary advantages and common development of each city have created tremendous economic growth momentum in the past decades.

These macro-economic conditions have provided a fertile soil for the rapid development of new energy vehicles in Guangdong, Hong Kong, and Macau. As far as Hong Kong is concerned, the number of new energy vehicles has grown rapidly over the past eight years, from less than 100 vehicles in 2010 to 11,345 vehicles in October 2018 [32]. At present, there are about 83 types of new energy vehicles that have been approved for driving in Hong Kong, including 58 private cars and 25 public and commercial vehicles.

An important reason for the rapid development of new energy vehicles in Guangdong, Hong Kong, and Macau is the provision of convenient and fast-charging devices and extensive infrastructure coverage. In Hong Kong, for example, CLP Holdings Limited and Hong Kong Telecommunications Co., Ltd. jointly established an equity joint venture, Smart Charge (HK) Limited, in August 2016 to actively promote one-stop smart charging services for new energy vehicles in Hong Kong. It covers a number of residential, office, and public facilities throughout the territory. The charging service can be electronically paid through Hong Kong Telecom's mobile wallet for the convenience of the public.

Tax relief is another important reason for the rapid development of new energy vehicles in the Greater Bay Area. Take Hong Kong as an example. In the past few years, the HKSAR government has been implementing a full reduction of the first registration tax on new energy vehicles. This policy provides great encouragement and support for citizens to purchase new energy vehicles. At the same time, the Hong Kong Special Administrative Region government has taken into account the sharp increase in the number of private cars. It decided that this policy would expire on March 31, 2018, and the upper limit would be reduced to HK$97,500. However, taxes on commercial new energy vehicles and motorcycles have been retained. These items will be reduced in March 31, 2021. On the other hand, the HKSAR government has introduced a "one-for-one" scheme, in which eligible existing car owners purchase new energy private cars and destroy their eligible old private cars at the same time. This will result in higher first-entry tax deductions. The highest amount is HK$250,000. This policy is expected to further promote the penetration rate of new energy electric vehicles in Hong Kong.

In addition to financial subsidies, the government has also introduced a series of tax incentives for consumers and car companies, including the purchase of pure electric vehicles, plug-in hybrid cars exempt from vehicle purchase tax, and the sale of new energy vehicles. The VAT rate for parts and components has been adjusted to 13%. In addition to cooperating with the introduction of subsidies, local governments have also used a series of non-financial incentives such as unlimited licenses to encourage consumer to purchase new energy vehicles.

Apart from private cars, new energy public operating vehicles, such as electric buses, can also be seen as one of the future development directions of Guangdong, Hong Kong, and Macau. The HKSAR government started to promote electric buses in 2012. At present, it has subsidized the franchised bus companies to purchase 36 single-decker electric buses and has already carried out trial operation on various routes in Hong Kong. According to the public information of the Guangdong provincial government, as of the end of April 2018, there were 56,000 pure electric operating vehicles in Guangdong Province; and in 2017, new energy vehicles accounted for more than 95% of the new and updated public buses in the province. The Guangdong Provincial Department of Transportation plans to implement complete bus electrification in the Pearl River Delta city by 2020. The booming electric buses have also provided a new engine for economic growth in the GBA.

5 A Positive Contribution in Sustainable Energy and Green Finance for the GBA

At the end of 2015, "green finance" was included in China's "13th Five-Year Plan," which clearly outlined the establishment of a green financial system, including the development of green credit, green bonds, and the establishment of a green development fund [33]. In August 2016, the seven ministries and commissions including the People's Bank of China issued the "Guiding Opinions on Building a Green Financial System," marking the official formation of China's national strategy for building a green financial policy framework. In June 2017, the State Council executive meeting decided to select places in Zhejiang, Jiangxi, Guangdong, Guizhou, and Xinjiang to develop a green financial reform and innovation pilot zone. The aim is being to conduct pilot exploration for the full implementation of China's green finance. The development of the green financial system officially entered the stage of implementation.

In August 2017, the central government approved the "Guiding Opinions on Building a Green Financial System," pointing out that the development of green finance is an important measure to achieve green development and an important part of supply-side structural reform. At the same time, they encouraged use of green credit, green bonds, green stock indices and related products, green development funds, green insurance, carbon finance, and other financial instruments and related policies to serve green development.

China actively promotes the issuance of green bonds in the international market and has caused the global green bond market to flourish. In 2016, the issuance of green-labeled bonds in the Chinese bond market exceeded 200 billion yuan. Accounting for nearly 40% of the global circulation, it has become the world's largest green bond market.

In the GBA, the macro-economic foundations of green finance have been well established. The GBA has consistently maintained a stable inflation rate in the past few years, much better than other regions of China. The education level of the population is generally high, and the unemployment rate has been at a low level for several consecutive years. In addition, its economic development has grown steadily, especially in innovative industries and services named as "new economic businesses." It is particularly noteworthy that the financial situation of many cities in the GBA is good enough to support the overall future development of the economy and can provide precious opportunities in the region.

In the past a few years, the GBA has successively promoted large-scale multi-dimensional cooperation platforms such as the districts of Nansha in Guangzhou, Hengqin in Zhuhai, and Qianhai in Shenzhen. The formation of new economic systems has been accelerating and is represented by innovative service industries. At the same time, the GBA has also effectively promoted the cooperation and development between the GBA and other mainland regions, especially for financial industries, providing convenient conditions for the economic development in both industry and capital markets.

In order to undertake research work in the area of green finance, there are great challenges in terms of data measurability and accountability.

At present, many scholars have emphasized the need to improve three aspects of green financial standards. First, there is a lack of consistency in standards for green bonds from the three rule makers—the Bank of China, the National Development and Reform Commission, and the exchange. Secondly, the standards of various green bonds are not completely consistent with the standards of green credit, resulting in assets identified as green credits not being able to match the green bond standards after securitization. Thirdly, domestic green bond standards are not completely consistent with international standards.

Therefore, we really need a new way to measure the "green value" of each economic activity to measure the degree of "greenness." Only when it can be accounted for and measurable, can it be recognized by the international community. It can then avoid regulatory arbitrage in domestic operations, so that the incentives and restraint measures discussed by all parties can be implemented.

The international research paradigm in dealing with global climate change is worth studying. Since natural science research shows that carbon dioxide is the largest greenhouse gas, carbon emission reduction is the primary goal. All economic activities therefore have a corresponding carbon emission level or carbon neutral level. This is used to measure the environmental impact.

It is difficult to rely only on commercial activities to reveal the variables that can be used to measure the degree of "greenness." Relying on economics or finance research is the correct path, yet requires enormous amount of work. It requires a degree of interdisciplinary integration.

Nowadays, the commercial costs of undertaking green finance business are very high. It is difficult to evaluate green measures. Thus, obtaining "green value" from the output will engender a lot of uncertainties and risks. Those risks may affect public welfare. It is necessary to identify the degree of environmental protection of each enterprise in a relatively transparent environment. This kind of financial reporting system will also provide convenience in disclosing information.

At present, China has become one of the world's largest green debt issuers, and the types of green financial products are quite complex. China's economy is shifting toward greener development, and governments at all levels are very active in institutional innovation for green development [34]. All of these provide a rare historical opportunity for our academic community to conduct original research on green finance.

In June 2017 Guangzhou government established a Green Finance Reform and Innovation Pilot Zone in Huadu District. Over the past year, the Municipal Financial Bureau and Huadu District have closely focused on four major areas, namely the experimental field of green financial reform and innovation; the demonstration zone of coordinated development of green finance and green industry; the new platform for cooperation and development of the GBA; and the assistance of the construction of the "Belt and Road." They aim to firmly grasp the main line of green financial services in terms of real economic development, to vigorously promote the innovation of green financial mechanisms, and to improve the ecological environment of green finance development. Up to now, the green financial zone has helped establish 151 green institutions with a registered capital of 6.9 billion yuan. Several world-class enterprises have set up pioneering projects in the zone, such as CLP and Nissan. They are developing new energy vehicles and intelligent network automobiles, with strong financial resources and support from strategic emerging industries. The green financial zone also actively promotes green travel and enables financial leasing companies to provide green financing services for one hundred electric buses under operation in Huadu district. The green financial zone strives to fully realize bus electrification in Huadu district in the near future and is considering the construction of green buildings during the process of regional development and urban renewal.

The development of green finance has long-term characteristics. As an institutional arrangement, the green financial system mainly supports economic greenification through financial credit, green bonds, green stock indices and related products, green development funds, green insurance, carbon finance, and other financial instruments. Its occurrence, development, and maturity will require a long process of gradual development.

6 Future Challenges

In future, the Great Bay Area of China is likely to strengthen its links with the mainland industry and achieve strategic goals in economic cooperation between Hong Kong and Macau and the Mainland. It is expected to maintain steady economic growth and to build a green and healthy sustainable development zone. At the same time, the GBA has a unique geographical location and certain legal advantages. For example, Hong Kong has a key role as an important international shipping center. It is a global economic exchange for "Belts and Roads" Initiatives, and a hub to "go global" for Chinese enterprises. The GBA will take a leading role in the "National Manufacturing 2025" and other national plans as well.

Challenge 1: Cost of production

In October 2017, UBS released a report on consumer acceptance of electric vehicles [35]. The report selected consumers from the world's major auto markets as the survey targets, and finally found that most consumers prefer cars with pure electric engines rather than "electric-fuel" dual mode. Concerns focus on high prices, limited cruising range, insufficient number of charging stations, and short battery life. Among these concerns, the high price is the primary reason why most people are not willing to buy electric cars.

A survey conducted by Goldman Sachs in 2016 also found that 50% of consumers believe that the high price is the main factor affecting the purchase of electric vehicles [36], followed by limited cruising range and charging facilities.

Electric vehicles appear as substitutes for fuel vehicles. Similar to the situation when fuel vehicles began to spread rapidly, the rapid decline in cost through technological advancement and efficiency improvement has become the primary factor affecting the popularity of electric vehicles.

Regardless of whether it is an electric car or a fuel car, it is composed of key elements such as a power system, a vehicle body, a chassis, an automotive electronic system, and interior and exterior parts; but the cost ratio of each part is not the same. For pure electric vehicles, the battery and powertrain systems account for up to 50% of the total cost, while for fuel vehicles, the engine and powertrain systems account for only 15% of the total cost. This part of the cost structure difference is the main reason why the current cost of electric vehicles is higher than that of fuel vehicles. How to reduce costs through technological advancement and scale effect is the primary problem that the industry needs to solve.

Challenge 2: Cruising range

Under the same external conditions such as vehicle weight and temperature, the cruising range of the electric vehicle is mainly determined by the battery capacity. The larger the battery capacity, the higher the cruising range. Under normal circumstances, the mileage of one-degree electricity is 6–7 km. Therefore, the electric vehicle needs to achieve a higher cruising range. The amount of energy that can be carried (i.e., energy density) can also increase overall battery capacity. The core of the former is the cost issue, and the core of the latter is the technical and material issues.

In those domestically sold pure electric vehicle in 2018s, the cruising range was about 300 km, which can only meet the needs of urban traffic or short-distance travel. A typical fuel vehicle, however, with 40–50 L of fuel tank capacity can usually run more than 500 km.

Challenges 3: Charging facilities

Although the number of charging facilities in China has increased, it is still unable to catch up with the rapid growth of electric vehicles on the road. At present, the main factors restricting the construction of charging facilities include long power access period, high cost, difficulty in installing measurement meters, tight urban land supply, and difficulties in site selection. In addition, private construction of charging facilities also faces some other problems, such as construction quality, reporting standards, and auditing issues. The filing process is extremely time-consuming. In terms of charging operation services, there are also technical challenges like interconnection of online payment system, switching between new and old systems, payment safety issues, etc. Currently, it is difficult for private enterprises to make profits through investing in charging stations.

Since some cities have established strict licensing regulation, allowing only new energy vehicles to be licensed free of charge, the drivers of electric vehicles are mainly concentrated in first- and second-tier cities. Those cities are also under vehicle purchase restrictions, such as Beijing and Shanghai. However, for these first-tier cities, there is a widespread shortage of parking spaces. Fixed parking spaces are often the prerequisite for the installation of charging piles in urban residential quarters. These could fulfill the charging needs of more than half of the electric vehicles in Beijing, Shanghai, and other cities. Owners still have to rely on public parking charging facilities.

As for the third- and fourth-tier cities, electric vehicles are currently sold less and are mainly low-speed vehicles. Most of these electric vehicles use lead–acid batteries as the power source. They do not support fast charging, and their electric vehicles generally have independent garages. Therefore, public charging piles or special charging piles built in third- and fourth-tier cities are not considered profitable at the current stage.

The development of the whole industrial chain for new energy vehicles can not only effectively utilize the technological innovation advantages of the GBA, but can also promote the employment and cultivation of high-tech talents. Infrastructure and transportation facilities are among the most important types of investment in economic exchanges and cooperation between countries and regions along the "Belts and Roads" Initiative. New energy vehicles could meet the requirements of such investment cooperation. At the same time, "Made in China 2025" also puts higher requirements on the production design capabilities of new energy vehicles. The innovative design and production capacity of new energy vehicles in the GBA has been greatly improved. Great progress in important areas such as new energy vehicle charging and discharging technology, battery safety management models, and a combination of electric vehicle and power grid technology need to be achieved in future.

The GBA will continuously attract high-tech talent to create scientific and technological values and realize the combination of production, education, and research.

From a technical point of view, another reason why new energy vehicles can be generally accepted by Hong Kong and Macau is that the land area of Hong Kong and Macau is not large, and the cruising range of new energy vehicles is sufficient to meet the daily needs of users in Hong Kong and Macau. With the development of innovative battery technology for new energy vehicles, the cruising range continues to increase. In addition, the charging facilities and related infrastructure in the GBA are well prepared. However, the standardization of charging devices and related infrastructures needs to be unified as much as possible in the future, in order to provide an important basis for the further popularization of new energy vehicles in Guangdong, Hong Kong, and Macau.

Finally, the GBA has a solid market economy and financial industry foundation. Future policy-makers may consider setting up an innovation development fund to encourage innovative high-tech industry research works. Such funds may be jointly established and managed by the government, new energy companies, and financial institutions. The government could encourage high-tech industry to innovate in science and technology, to combine production, education, and research by injecting seed funds. All those suggestions are closely related to talent and intelligence exchanges. The GBA could also consider strengthening exchanges between the various places, encouraging Hong Kong and Macau to invest in Guangdong, and encourage high-tech people from the Mainland to study and do exchanges in Hong Kong and Macau. Mutual investment would therefore be established. By exploring the development space of start-ups and small-to-micro enterprises, youth entrepreneurship and employment rate in the GBA would be largely promoted.

References

1. Emadi A, Rajashekara K, Williamson SS, Lukic SM (2005) Topological overview of hybrid electric and fuel cell vehicular power system architectures and configurations. IEEE Trans Veh Technol 54
2. Huo H, Zhang Q, Wang MQ, Streets DG, He K (2010) Environmental implication of electric vehicles in China. Environ Sci Technol 44:4856–4861
3. Li X, Clark CD, Jensen KL, Yen ST, English BC (2013) Consumer purchase intentions for flexible-fuel and hybrid-electric vehicles. Transp Res Part D: Transp Environ 18:9–15
4. Geng Y, Ma Z, Xue B, Ren W, Liu Z, Fujita T (2013) Co-benefit evaluation for urban public transportation sector—a case of Shenyang China. J Clean Prod 58:82–91
5. Sato Y, Ishikawa S, Okubo T, Abe M (2011) Development of high response motor and inverter system for the Nissan LEAF electric vehicle. SAE Technical Paper 2011-01-0350
6. Endsley MR (2017) Autonomous driving systems: a preliminary naturalistic study of the tesla model S. J Cogn Eng Decis Making
7. Khaligh A, Li Z (2010) Battery, ultra-capacitor, fuel cell, and hybrid energy storage systems for electric, hybrid electric, fuel cell, and plug-in hybrid electric vehicles: state of the art. IEEE Trans Veh Technol 59
8. Office of Low Emission Vehicles (2013) A strategy for ultra low emission vehicles in the UK. The UK Government
9. Department of Transportation (2017) Improving air quality in the UK: tackling nitrogen dioxide in our towns and cities. The UK Government. 2017
10. Department of Business, Energy, and Industrial Strategy (2017) Clean growth strategy. The UK Government
11. The International Council on Clean Transportation (2018) The continued transition to electric vehicles in US cities. Washington DC US
12. Sovacool BK (2017) Experts, theories, and electric mobility transitions: toward an integrated conceptual framework for the adoption of electric vehicles. Energy Res Soc Sci 27:78–95
13. Rezvani Z, Jansson J, Bodin J (2015) Advances in consumer electric vehicle adoption research: a review and research agenda. Transp Res Part D 34:122–136
14. Hensley R, Knupfer SM, Krieger A (2011) The fast lane to the adoption of electric cars. McKinsey Quarterly
15. Gong H, Wang MQ, Wang H (2013) New energy vehicles in China: policies, demonstration, and progress. Mitig Adapt Strat Glob Change 18:207–228
16. Wang Y, Chang X, Chen Z, Zhong Y, Fan T (2014) Impact of subsidy policies on recycling and remanufacturing using system dynamics methodology: a case of auto parts in China. J Clean Prod 74:161–171
17. Wan Z, Sperling D, Wang Y (2015) China's electric car frustrations. Transp Res Part D: Transp Environ 34:116–121
18. Zhang X, Wang K, Hao Y, Fan Jing-Li, Wei Yi-Ming (2013) The impact of government policy on preference for NEVs: the evidence from China. Energy Policy 61:382–393
19. Du J, Ouyang M, Chen J (2017) Prospects for Chinese electric vehicle technologies in 2016–2020: ambition and rationality. Energy 120:584–596

20. Ou S, Lin Z, Qi L, Li J, He X, Przesmitzki S (2018) The dual-credit policy: quantifying the policy impact on plug-in electric vehicle sales and industry profits in China. Energy Policy 121:597–610
21. Ministry of Finance of PR China (2016) Notice on the liquidation of new energy vehicle subsidy funds
22. Ministry of Industry and Information Technology of PR China (2017) Action plan for promoting the development of automotive power battery industry
23. Ministry of Industry and Information Technology of PR China (2018) Guide to the construction of national vehicle network industry standard system
24. Ministry of Industry and Information Technology of PR China (2017) Medium- and long-term development plan for the automobile industry
25. Ministry of Industry and Information Technology of PR China (2017) Regulations on the administration of new energy vehicle production enterprises and products
26. National Energy Administration of PR China (2017) Guidance on promoting energy storage technology and industrial development
27. State Administration of Taxation of PR China (2018) PR China vehicle purchase tax law
28. China Electric Vehicle Charging Infrastructure Promotion Alliance (2018) Annual Report of China's Electric Vehicle Charging Infrastructure Development in 2017. Power Division, National Energy Administration
29. Hong Kong Trade Development Council (2017) Statistics of the great bay area
30. China Southern Power Grid (2018) Monthly Press Release
31. Kwok RH (2018) Guangdong New Energy operates 61,000 vehicles. The intercity charging network system was completed in 2020. Ta Kung Pao
32. HKSAR Environmental Protection Department (2018) Promoting the use of electric vehicles in Hong Kong
33. Ng S, Mabey N, Gaventa J (2016) Pulling ahead on clean technology: China's 13th five year plan challenges Europe's low carbon competitiveness. E3G. Bloomberg
34. Wang Y, Zhang R (2017) China's Green Bond Market. Int Capital Market Features (44). Q1
35. Ganter R, Berrisford C, Dennean K, Dessloch S (2017) Longer term investments—smart mobility. Chief Investment Office, UBS
36. Yuzawa K, Burgstaller S, Yang Y, Tamberrino D (2016) Electric vehicle boom: ice-ing the combustion engine. Equity Res Goldman Sachs (2016)

Analysis of Economic Benefits of the Energy Internet and Its Facilitation of Green Development

Jingyan Fu and Shuyao Wang

Abstract The development of the Energy Internet Demonstration Project and the construction of the smart grid in the Bay Area of Guangdong, Hong Kong and Macao are to induce measurable economic benefits for the green development of the Greater Bay Area. Based on the structural characteristics of the Energy Internet, this paper analyses the development advantages of the smart grid and the current status of energy development in the Greater Bay Area. Focusing on the Energy Internet's energy saving, industrial transformation and environmental economy, key economic benefits for the Greater Bay Area's green development can be summarized as follows. First, the Energy Internet can reduce outages and electricity tariffs to achieve cost savings. Second, it can facilitate the transformation of traditional industries and the development of new industries while realizing the benefits of industrial upgrades. Third, the Energy Internet can create environmental and economic benefits through facilitating high value-added resource conservation, pollution control fund savings as well as mitigating measures in response to global climate change. To a significant extent, the technological innovation within the energy sector has positive policy implications for sustainable development of the Greater Bay Area.

Keywords Energy internet · Smart grid · Economic benefits

1 Introduction

In the face of increasing resource and environmental problems, governments have to consider massive reform of the energy system. American scholar Jeremy Rifkin proposed the "Energy Internet" in his book *The Third Industrial Revolution* (2011). The "Energy Internet" is a distributed, open and shared network based on renewable energy, to better cope with issues such as excess energy, inefficiency and environmental pollution. Zhihong and Jian [1], Pourbabak and Chen [2] elaborated on the operation mode of the Energy Internet's different energy networks in their research,

J. Fu (✉) · S. Wang
College of Economics, Institute of Resource, Environment and Sustainable Development, Jinan University, Guangzhou, China
e-mail: tfjyan@jnu.edu.cn

J. Fu and A. W. Ng (eds.), *Sustainable Energy and Green Finance for a Low-carbon Economy*, https://doi.org/10.1007/978-3-030-35411-4_7

fully demonstrating that the Energy Internet is the only way for future development; Zhou and Yang [3] and others conducted in-depth analysis of the global Energy Internet, analysing its specific characteristics and advantages. In terms of technology cost, Si and Wang [4] proposed multi-energy management (distributed energy network) energy complementary strategy, which can achieve the maximum total cost reduction at the compromise solution point and effectively reduce energy costs. These reduced costs would achieve a balance between supply and demand while improving social and economic benefits. Many scholars such as Qi et al. [5] have studied the international influence of energy interconnection based on a global perspective and laid a solid foundation for the development of the Energy Internet.

On 26 September 2015, President Xi Jinping announced at the United Nations Development Summit that "China is proposing to explore the construction of a global Energy Internet to promote a global demand for electricity in a clean and green way". At the same time, the State Council issued the "Guiding Opinions on Actively Promoting the "Internet+" Action", which proposed the concept of "Internet + Smart Energy". In 2016, the 13th Five-Year Plan specifically proposed "constructing the development of source-network-load-storage" and integrating and complementing the Energy Internet. China's Energy Internet began to formally enter the preparatory stage and released "Promoting "Internet+" Guiding Opinions on the Development of Smart Energy". This guidance will promote the development of the Energy Internet in two phases, pilot demonstrations in the early stage and promotion and application in the later stage to ensure effective results.

The Energy Internet has also greatly promoted the future development of Guangdong, Hong Kong and Macau. On 18 March 2019, China Southern Power Grid announced that it will accelerate the planning and construction of smart grids in the Bay Area in the next five years. Energy Internet demonstration projects such as Guangzhou, Shenzhen and Zhuhai are also actively under construction. It is estimated that the Greater Bay Area power load will reach 1.3 billion kilowatts in 2035, which requires sufficient power support from the Energy Internet. In the development of the Bay Area, Guangzhou and Shenzhen, the key industries of science and technology innovation and large-scale manufacturing have a large electricity demand and need the Energy Internet to help provide stable transmission of electricity. At the same time, the Energy Internet has also promoted energy transformation and upgrades. The use of renewable energy to generate electricity that drives industrial transformation and upgrade has environmental and economic benefits for low-carbon green development of the Bay Area.

Although China has made major breakthroughs in Energy Internet and smart grid construction in the Greater Bay Area, the current theoretical foundations for smart grid construction in the Bay Area are weak. There are also many technical and application-level problems such as the Guangdong-Hong Kong-Macao cross-region. The technical difficulties of transmission and the contradiction between the imbalance of electricity prices of industrial electricity and residential electricity use need further research and analysis to achieve balanced and coordinated interregional development.

2 Overview of Energy Internet and Smart Grid Construction in the Greater Bay Area

2.1 Energy Internet Concept and Its Mode of Operation

The Energy Internet is based on the power system. The UHV grid is the backbone grid, driven by clean energy. It is based on power electronics and ICT systems, decentralized energy storage and energy transmission for centralized or distributed renewable energy. In addition to the consumption load, an energy allocation platform is formed to achieve tight coupling and coordinated sharing of energy and information.

As shown in Fig. 1, its mode of operation mainly uses traditional power generation methods such as wind, photovoltaic, bio-power and others to provide energy sources and energy big data through a distributed intelligent energy management system (IEMS). Statistically, the energy of each region is rationally distributed through the backbone network and the local area network, and finally, the industrial routers connect industrial, commercial and residential buildings, and energy consumers can exchange energy and break its one-way flow. The traditional model generates capital flows, and the Energy Internet's information flow controls the flow of energy to ensure the security and reliability of the entire system. At the same time, the Energy Internet uses the large power grid as the "main trunk network" and the "micro-network" as the local area network, so that it can receive the input of new energy to the maximum extent and output the energy more widely.

Electric energy has great advantages in energy transmission efficiency, so it will be the core content of future energy interconnection construction. The large power grid has an unparalleled advantage in terms of transmission efficiency and bears the

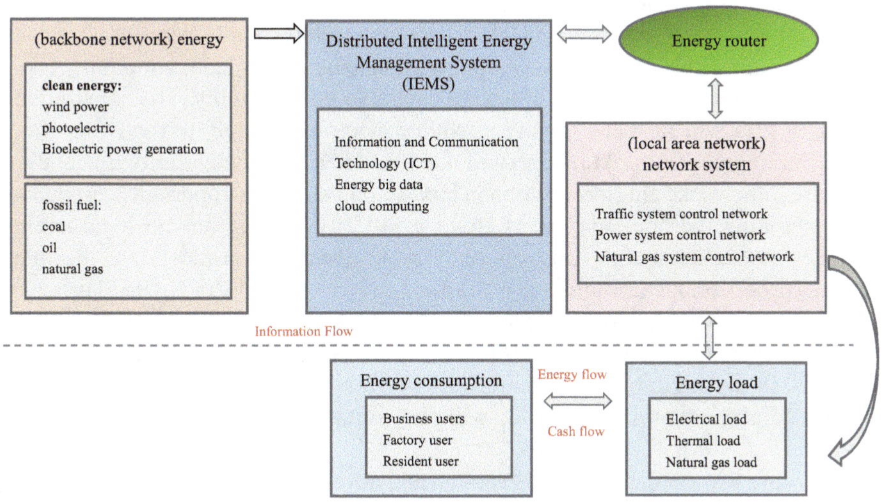

Fig. 1 Overall structure of the Energy Internet. *Source* Author draws

function of the "backbone network"; the micro-grid is a key component of the Energy Internet is formed through the distribution of local electricity area networks. This enables the Energy Internet to transform the centralized and one-way grid into an interactive communication platform for energy, achieving overall optimization and real-time adjustment of energy production, transmission and distribution and greatly improving the efficiency of energy development and utilization.

2.2 Greater Bay Area Smart Grid Construction

"First-class construction of the Greater Bay Area requires first-class power security". On 18 March 2019, China Southern Power Grid issued the "Key Measures for Serving Bay Area of Guangdong, Hong Kong and Macao Development" and proposed 26 specific measures to serve the development of the Greater Bay Area. In 2022, a safe, reliable, green and efficient smart grid basically will be built; in 2030, a world-class smart grid will be built.

At present, the current Greater Bay Area power supply situation makes smart grid construction imperative. Since the mainland and Hong Kong and Macao power grid interconnection, China Southern Power Grid has gradually increased the proportion of power supply to Hong Kong and Macao. It is an important subject for mutual supply of electricity from Guangdong, Hong Kong and Macao. According to its data, China Southern Power Grid's electricity consumption for Hong Kong accounts for approximately 25% of its total electricity consumption, reducing carbon dioxide emissions by approximately 27 million tons; electricity consumption to Macao accounts for 70% of its total electricity consumption, and it transmits electricity to Macau. The third channel is under construction; the southern region has built an "Eight Crossing and Direct" west-to-east power transmission channel, and the utilization rates of wind and photovoltaic power generation are all over 99%. The four large Bay Area cities of Foshan, Dongguan, Shenzhen and Guangzhou have taken over the mainland. The top four cities have reliable power supply. The construction of a smart grid will further promote a reliable electricity supply to the Hong Kong and Macao regions in the Mainland and meet the Bay Area's electricity demand.

Figure 2 shows the energy situation in Hong Kong and is an important prerequisite for accelerating the construction of smart grids. Hong Kong has no local energy resources and relies mainly on imports. Among them, oil products are the main imported products and the mainland is the main source of supply. In 2018, oil products accounted for 75.6% of imports and electricity imports accounted for only 3.3%. However, in the final energy demand, electricity demand accounted for approximately 48% in 2018, which implies a potential cap on Hong Kong's electricity supply. It is envisaged that smart grid construction will play a vital role in Hong Kong's electricity supply.

According to the smart grid construction and development plan, the power supply will be greatly satisfied. It is estimated that by 2035, the electricity load of the Greater Bay Area of Guangdong, Hong Kong and Macao will reach 130 million kilowatts, the

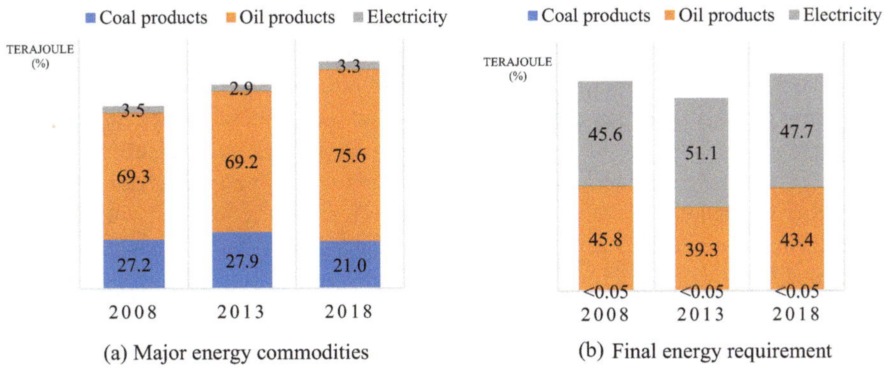

Fig. 2 Hong Kong's energy imports and consumption. *Source* https://www.censtatd.gov.hk/home.html

total electricity consumption of the whole society will reach 700 billion kWh, and the per capita electricity consumption will exceed 8300 kWh, which is equivalent to the sum of electricity consumption of Hong Kong, Macao and Taiwan. This combined with renewable energy generation and the national Energy Internet should lead to enhancement of energy interconnection in the Greater Bay Area.

2.3 *Analysis of the Advantages of Developing Energy Internet in the Greater Bay Area*

The core of the Energy Internet is to build a network platform that integrates information flow, energy flow and capital flow. Until now, pilot projects such as Guangzhou, Zhuhai and Shenzhen have been gradually carried out and the smart grid project in the Bay Area has been launched. The complementary advantages between the Energy Internet and the Greater Bay Area have also begun to gradually unveil.

First, the large number of information exchange bases is conducive to accelerating the process of power marketization. Major manufacturing enterprises and technological innovation enterprises are gathered within the Greater Bay Area, and electricity demand is large; therefore, the information exchange base is huge. Individuals and enterprises can grasp their own energy use through network platforms, make reasonable sales decisions on surplus energy and promote the marketization of electricity prices. The government only needs to play a supporting role to avoid regional electricity price imbalance caused by government pricing, so that it maximizes economic benefits. At the same time, power generation companies can query electricity consumption in real time, adjust production in real time, avoid overcapacity and strengthen their control of the power supply in the electricity market.

Second, the two-way promotion of the Bay Area and the Energy Internet are developed together. "Ecological priority, green development" is the basic principle for the development of the Bay Area of Guangdong, Hong Kong and Macao—it strives to develop clean energy to replace traditional fossil energy. To increase the power supply of clean energy in the Bay Area, it is estimated that by 2035, the proportion of installed clean energy capacity in the Bay Area will reach 80%, which will promote the clean power generation capacity of the Energy Internet to a certain extent. The Energy Internet can replace traditional fossil energy such as coal with clean energy such as photovoltaics, which can effectively reduce the emissions of carbon dioxide and related atmospheric pollutants and improve the Bay Area's energy structure.

Third, power stability helps develop the Bay Area industry. Traditional energy sources use concentrated production and large-scale transmission to deliver energy to end consumers. Power failures such as outages and outages caused by unstable circuits are huge. The Energy Internet uses a centralized and distributed energy network system to support large-scale access to distributed generation and storage systems. Due to the transmission and distribution efficiency advantages of "backbone network" and "local area network", the Energy Internet can easily connect large-capacity energy storage devices and stable energy output, which helps large-scale stable power consumption of Bay Area industry and avoids industrial losses due to unstable circuits.

3 Facilitating Green Economic Development in the Greater Bay Area

In the development of the Greater Bay Area, electricity infrastructure is considered as a salient area for sustainability planning. In supporting the Energy Internet development, China Southern Power Grid Corporation announced the acceleration of smart grid construction in Guangdong, Hong Kong and Macau, which means that "Internet + Smart Energy" is gradually on the right track in the Greater Bay Area development plan. The Energy Internet boosts the Greater Bay Area's green development mainly in terms of cost savings, industrial upgrading and environmental economics.

Figure 3 is a schematic diagram of how the Energy Internet's green development mechanism boosts the Bay Area. From the perspective of power generation cost, the Energy Internet solves the problem of power loss caused by circuit instability due to long-distance transmission and improves power supply capacity. The power price decline caused by electricity price benefits further strengthens power use and supply capacity. In the list of the first batch of "Internet+" smart energy (Energy Internet) demonstration projects announced in March 2017, there are five projects in the Greater Bay Area cities, including Zhuhai, Guangdong Province, which supports the energy consumption revolution. The "+ Smart Energy Demonstration Project" passed the acceptance test and is the first project to do so. Cities such as Guangdong, Shenzhen and Zhuhai have become the forerunners of power development in the Bay Area, providing strong support for its power supply, meeting the electricity

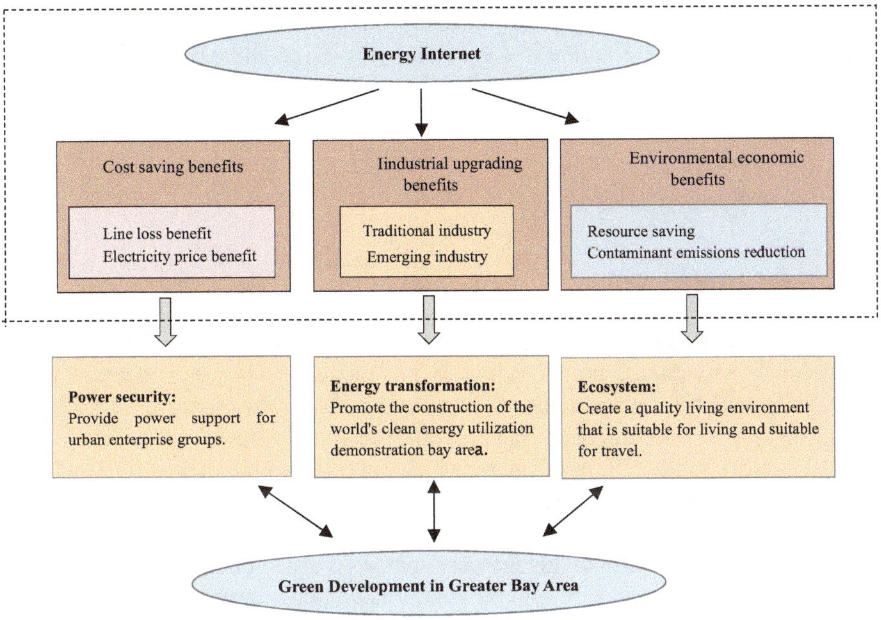

Fig. 3 Energy Internet boosts the Greater Bay Area's green development mechanism. *Source* Author draws

demand of the Greater Bay Area's large manufacturing and innovative industries and requirements for power stability. The level boosts the rapid and quality development of the Greater Bay Area's economy.

The Energy Internet's gradual replacement of fossil energy with renewable energy has also contributed to the adjustment and upgrading of industrial structure. On the one hand, it has intensified the overcapacity of the traditional fossil energy industry and stimulated its technological innovation; on the other hand, the use of renewable energy has driven the development of a number of emerging industries, especially electric vehicles. At the same time, the Greater Bay Area is home to various innovative enterprises and technology industries and has great potential and capabilities in industrial innovation. In addition, China Southern Power Grid proposed the construction of a smart grid in the Bay Area. It plans to invest more than 170 billion yuan between 2018 and 2022, which will stimulate the vigorous development of the renewable energy industry and help the Bay Area become a dynamic and internationally competitive first-class Bay Area.

From the perspective of environmental economic benefits, the Energy Internet has saved high value-added resources such as land. Especially for the Greater Bay Area, where land is expensive, the environmental economic benefits of resource conservation are particularly obvious. Second, Energy Internet construction uses new energy to generate electricity, reduces emissions of carbon dioxide and its coordinated pollutants, reduces pollution prevention costs and governance funds, and better copes with the challenges brought about by climate change. This helps achieve the goal of building a "quality living circle that is suitable for living and suitable for travel" in the Greater Bay Area.

3.1 Cost-Saving Benefits

3.1.1 Line Loss Benefit

The distributed energy network transmission mode enhances power stability and can reduce power loss during transmission, that is, generate line loss benefits. China's traditional power generation mode is centralized power generation and long-distance transmission, and long-distance transmission must lead to power loss. The degree of loss depends on the length of the transmission line and the degree of congestion in the transmission path. The relevant data show that the line loss can reach 20% during the peak power consumption period. The geographic location of the Energy Internet is distributed close to the end user, which greatly reduces power loss during transmission. As shown in Fig. 4, in the traditional mode, power loss of high-voltage power from the transmission network to the load end can reach the entire line. After distributed energy access, the power is stepped down, power loss is gradually reduced, and the transmission stability is gradually enhanced. Referring to the distributed power generation economic benefit analysis and its evaluation model (Ming 2010), the line loss model can be used to reasonably design the distributed energy access and operation mode, which can effectively reduce the system line loss.

For the development of the Bay Area, the distributed grid enhances the stability of the circuit and effectively reduces the average power outage time for urban customers. In its "26 key measures", China Southern Power Grid estimated that the annual power outage time of central customers in Guangzhou and Shenzhen will be less than 0.5 h

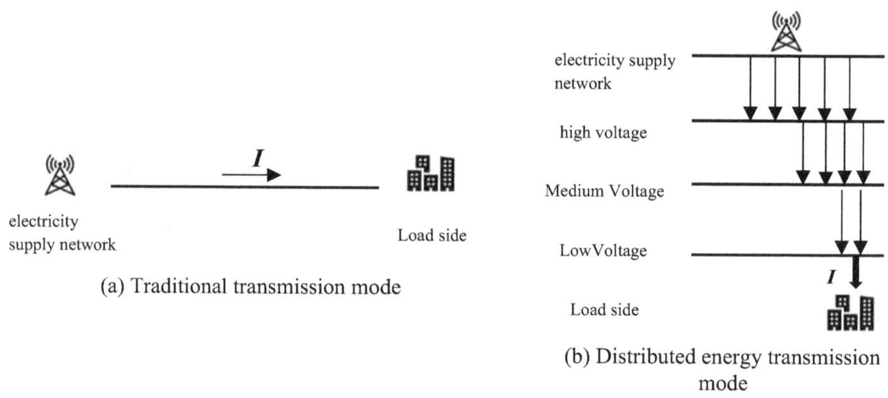

Fig. 4 Energy transmission mode. *Source* Authors

per year in 2020, for the average annual customer base in Zhuhai, Foshan, Huizhou, Dongguan, Zhongshan, Jiangmen and Zhaoqing. The power outage time is less than 1 h for the high-reliability power supply demonstration zones such as Guangzhou Zhongxin Knowledge City, Shenzhen Qianhai and Zhuhai Hengqin. A new district will be built. The average annual power outage time is less than 5 min, reaching the leading level of similar international cities. First-class power service will effectively protect the first-class construction of the Greater Bay Area and create a first-class social life order.

3.1.2 Electricity Price Benefits

The decline in power generation costs can be considered from two perspectives. On the one hand, construction of the Energy Internet infrastructure and minimization of operating costs have led to a continuous decline in average costs. On the other hand, renewable energy itself has the advantage of generating electricity costs and is widely used in the Energy Internet. After the promotion, the cost advantage naturally turns into an economic advantage.

Average power generation cost

The total investment of Energy Internet construction can be divided into capital cost and operating cost. The cost of capital is the one-time investment in the Energy Internet infrastructure project after implementation of relevant policies in 2016, including the initial project construction costs. The operating cost is the marginal cost of using energy such as wind and solar after the initial construction of the Energy Internet project, including operation and maintenance costs and loan interest. Due to the Energy Internet project's long recovery period, the average cost after amortizing to the annual capital recovery cost is declining. Using the project payback period method to calculate electricity price, the average power generation cost is declining; that is, the electricity price will be on a decreasing trend. Taking distributed photovoltaic power generation in the context of the Energy Internet as an example, and referring to the calculation of distributed photovoltaic power generation cost by scholars such as Zou Yuxin and Shi Quansheng, the following formula group can be established.

Advantages of renewable energy generation

According to the International Energy Agency's "Global Energy Transformation: 2050 Roadmap" issued in April 2018, the total amount of renewable energy must rise from 18% of today's energy consumption to 67% in 2050. In addition, the application of renewable energy in the power sector should rise from 25% today to 85%, mainly from wind power and photovoltaic power generation. This requirement has the same internal appeal as the Energy Internet project, which has vigorously promoted the technological development of renewable energy utilization. The renewable energy cost advantage also promotes the economic benefits of the Energy Internet. Although renewable energy currently is in the early application stage, the cost advantage of power generation is not obvious; but according to the learning curve theory and the expansion of scale benefits, its cost advantage will be obvious in the next few years. In recent years, data show that renewable energy has entered a virtuous cycle of cost reduction and accelerated technological progress.

Figure 5 shows the 2010–2017 renewable energy weighted average generation cost (LCOE, taking into account investment, operating costs and benefits over the life cycle, including capital costs), calculated at 2016 constant dollar prices. The blue dot indicates the weighted international average LCOE level. Except for the slight increase in geothermal power generation and hydropower cost, the cost of renewable energy power generation continues to decline, especially in photovoltaic power generation, with an average of 0.36 USD/kWh in 2010, to an average price of 0.1 USD/kWh in 2017, a decline of 72%. Comparatively, the LCOE cost of fossil energy power generation in 2017 is approximately 0.05–0.17 USD/kWh, and it can be seen that renewable energy power generation cost has gradually fallen below or close to fossil energy power generation. Wind power projects also have great potential for future cost advantages. The main limitation of wind power projects lies in the cost disadvantage of "West-East Power Transmission" and the transmission restrictions of power grids. Construction of the Energy Internet can solve the problem of grid transmission and imbalance of power area, which has great potential for future power generation cost savings.

Whether or not companies choose to set up bases or subsidiaries in the Bay Area, electricity prices are an indispensable consideration. Most of the Greater Bay Area industries are high tech involving a huge amount of information and modern manufacturing, which are often high-energy industries. Silicon Valley in the San Francisco Bay Area is home to a large number of high-tech companies, and data centres such as Microsoft and Google can generate electricity from medium-sized power plants. The Mitsubishi and Matsushita factories in the Tokyo Bay area also use a large amount of electricity. Therefore, the electricity price in the Bay Area is a crucial choice for the company, which also affects the sustainable development of the Bay Area.

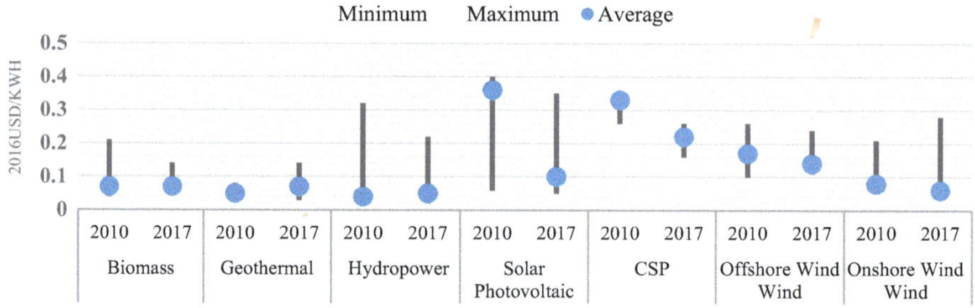

Fig. 5 Renewable energy weighted average power generation costs for 2010–2017. *Source* https://www.irena.org

3.2 Industry Upgrade Benefits

The Energy Internet boosts the energy transformation and upgrade of the Greater Bay Area mainly through the impact on traditional and emerging industries. The crisis of traditional industries facing elimination due to high pollution has forced them to carry out energy technology transformation; emerging industries have taken the lead in demonstration and effect in the Bay Area due to prosperous development leading the advancement of industrial technology.

Transformation and upgrading of traditional industries

The traditional industries are mainly labour-intensive dominated by processing and manufacturing industries. The Pearl River Delta region is a gathering place for traditional manufacturing industries. In the context of Greater Bay Area development, traditional manufacturing industries have exposed the disadvantages of insufficient vitality and high pollution. The impact of the Energy Internet's development on the transformation and upgrade of traditional enterprises in Guangdong, Hong Kong and Macao is mainly reflected in the energy substitution of highly polluting enterprises.

Presently, the Greater Bay Area is still based on fossil energy consumption, and the proportion of clean energy is not high. It is mainly because of the coal and oil cost advantage that high-polluting enterprises still use traditional energy production to maximize profits. According to the 2020 energy consumption forecast data of some major high energy consumption projects in Guangdong, it can be seen that high-energy-consuming industries have a greater impact on energy consumption, and this consumption has an impact on the "13th Five-Year" energy-saving target by as much as 52%. This explains that it is imperative to save energy and upgrade high-energy enterprises.

The large-scale application of renewable energy within the Greater Bay Area has forced high-slung companies to abandon traditional energy and switch to clean and renewable energy. The Bay Area can take advantage of favourable coastal geographic advantages, make full use of marine resources, develop offshore wind power, wave energy, tidal energy, etc., and further expand the use of Energy Internet clean energy. According to China Southern Power Grid data, in 2018, it sent 217.5 billion kWh of electricity, a year-on-year increase of 7.2%, and the utilization rate of water and electricity was 95.6%. The utilization rate of wind and photovoltaic power generation exceeded 99%. The net non-fossil energy accounted for 51.5% of electricity. With the construction of smart grids, the proportion of clean energy power generation will continue to rise, and eventually, the replacement of fossil energy will become the main mode of energy use.

On the other hand, the incentives and pressures of traditional industries can lead to transformation and upgrading. The construction of the Energy Internet and the smart grid in the Bay Area will greatly develop energy transmission technology. Through the energy-saving mode of contract energy management, energy-saving costs will reduce enterprise costs and encourage traditional enterprises to save energy and reduce emissions. At the same time, according to the "Opinions on Innovation and

Improvement of the Green Development Price Mechanism" issued by the National Development and Reform Commission in June 2018, the unreasonable price preferential policies for high-energy-consuming industries will be completely eliminated, as will the elimination and restriction categories. Enterprises use electricity to implement higher prices. High-soil enterprises will face greater pressure to achieve transformation and upgrades.

Development of emerging industries

Construction of the Energy Internet will promote the development of a series of emerging industry sectors, which will lead to the overall upgrading of Bay Area industry. For example, by 2030, electric vehicles will be used on a large scale. It is estimated that there will be 100 million electric vehicles in China. At the same time, with obvious peak load conditions, the Energy Internet can effectively avoid peak load and promote its further development. On the one hand, the Energy Internet can form a two-way interaction with electric vehicles. That is, electric vehicles can use power from the distributed power grid or return the surplus power to the power grid. The two form a strong coupling relationship, effectively avoiding large-scale charging to the power grid. It can also enhance the reliability and economy of power transmission and ensure its safe and economic operation.

Shenzhen is the best city in the country for the promotion and application of new energy vehicles. It is also the first city in the world where buses and taxis are basically electrified. As of 2017, the number of electric vehicles in Shenzhen is approximately 80,800 as of 20 January 2018. The monthly sales growth rate ranked first, reaching 29,048, leading electric vehicle industry development in the Greater Bay Area. In the future, it will radiate to the surrounding cities and drive the entire industry's development. The "Guangdong, Hong Kong, Macao and Dawan District Development Plan" also proposed "to vigorously develop smart transportation, smart energy", cultivating and expanding new energy, energy conservation and environmental protection, new energy vehicles and other industries, forming an industry centred on energy conservation and environmental protection technology research and development with a headquarters' base. This is a gathering belt, putting forward the practical requirements for the development of the Energy Internet in the Bay Area.

In addition, from 2018 to 2022, China Southern Power Grid will invest more than 170 billion yuan in the Greater Bay Area power grid, which will drive the development along the smart grid industry chain. From micro-grid and energy storage to cloud computing, a number of emerging industries are facing greater opportunities to inject new development momentum into Bay Area industry.

3.3 *Environmental-Economic Benefits*

The Energy Internet promotes the use of renewable energy and has a huge positive effect on the ecological environment, mainly in terms of resource conservation, conventional pollutant emission reduction and climate change.

First, the development of the Energy Internet will greatly reduce the use of fossil energy and generate savings from it. For the development of the Greater Bay Area, the distributed energy design reduces the consumption of high value-added land resources. Land resources in the Bay Area are tight, and the total area of cultivated land is small, approximately 7820 km^2. The construction cost of large, conventional coal-fired power plants is no longer environmentally economical; therefore, the advantages of developing energy interconnection in the Greater Bay Area with clean and renewable sources are particularly obvious, and the resource-saving effects can also be more significant, resulting in economic benefits of resource conservation.

Second, conventional pollutants are reduced. Conventional pollutant abatement effects generally refer to reductions in emissions of sulphur dioxide, nitrogen oxides and fine particulate matter. Through analysing the factors affecting air pollution in Guangdong, Hong Kong and Macao, the main source of air pollution in the Pearl River Delta region is the total energy consumption by various traditional manufacturing and production industries. The source of pollutants in Hong Kong and Macao is mainly driven by population growth, growth in overall energy consumption by inhabitants and industrial electricity consumptions. At the same time, it is found that environmental governance funds are associated with pollutants; therefore, developing the Energy Internet can alleviate pollution from fossil energy consumption in the Greater Bay Area, while saving a large amount of environmental management funds and generating good environmental and economic benefits.

Third, it will be a better response to climate change. The power industry is the main CO_2 emitting industry. According to statistics, China's firepower CO_2 emissions account for 40–50% of the total emissions. Constructing the Energy Internet effectively improves the use of clean energy and energy efficiency to achieve the emission reduction goal to a large extent.

4 Prospects for the Development of the Energy Internet in the Greater Bay Area

By analysing the economic benefits of the Energy Internet for green development of the Bay Area, it can be seen that the Energy Internet will reduce electricity consumption cost in the Bay Area, drive industrial upgrades to achieve those benefits, while conserving resources to reduce pollution control cost and promote the Greater Bay Area's green development process. Looking to the future, the Energy Internet is expected to enhance the development of an efficient power market, industrial upgrades and ecological environment as deliberated in the following three main areas:

(i) It facilitates the development of the power market and creates one that is market-driven with government assistance. As a location in the forefront of development, the Bay Area has great innovation flexibility. The gathering of high-quality talent better implements the emerging policies and plays a role. The construction of smart grids lays a solid foundation for energy interconnection, and the power is connected to the Internet. In the interconnection between supply and consumer, the market can play its pricing mechanism, break the monopoly of the power market, maximize the benefits and solve the problem of large energy consumption and high cost in the Greater Bay Area.

(ii) The industry in the Greater Bay Area is poised to upgrade themselves for the creation of a first-class high-tech smart bay area. The Energy Internet has been transformed by traditional industries. The development of emerging industries has promoted the comprehensive renewal of the Bay Area industrial chain and stimulated innovation and practice. The future Bay Area will be built into a world-class emerging industry, advanced manufacturing industry and modern service industry base, ranking among world-class cities. At the same time, the use of new energy has increased in the Greater Bay Area, and industries such as energy conservation, environmental protection, and new energy vehicles have grown stronger, demonstrating the role of domestic enterprises to promote the country's economic transformation.

(iii) It establishes a quality living circle that is suitable for living and sustainable tourism. The Energy Internet is expected to greatly enhance the Greater Bay Area's ecology. The large-scale use of new energy sources will reduce carbon emissions, meet the international response to climate change requirements and reduce synergistic pollutant emissions and improve air pollution. Water pollution will help the Greater Bay Area build a good living environment, and it will also attract high-quality talent to start business and form a virtuous circle of sustainable development in the Greater Bay Area.

References

1. Zhihong J, Jian H (2018) Energy internet—a new driving force for sustainable urban development. Energy Procedia 152:1206–1211
2. Pourbabak H, Chen T (2019) Centralized, decentralized, and distributed control for energy internet. The Energy Internet 3–19
3. Zhou K, Yang S (2016) Energy internet: the business perspective. Appl Energy 178:212–222
4. Si F, Wang Jinkuan (2018) Cost-efficient multi-energy management with flexible complementarity strategy for energy internet. Appl Energy 231:803–815
5. Qi T, Winchester N, Kapplus V et al (2013) The energy and economic impacts of expanding international emissions trading. MIT Joint Program on the Science and Policy of Global Change, Cambridge, Massachusetts

Pioneering Energy Efficiency Through Green Financing in Guangdong: Project Partnership with Multilateral Bank

Junyan Liu and Jingyan Fu

Abstract Green development has become the national strategy of China, and green credit is an important source of incentive for green development. The Guangdong Energy Efficiency Power Project (EEPP) is the first pilot project of this kind based on a partnership between the Chinese government and the Asian Development Bank, which is considered to be successful, in terms of both operations and environmental effects. Based on the analysis of this EEPP case, this study provides a set of relevant experiences for the development of green credit and green financing that can be potentially extended nationwide referencing on the Guangdong model for the sustainable development in China.

Keywords Green credit · Efficiency power plant · Energy-saving and emission reduction · Revolving fund

1 Introduction

Green finance is a phenomenon that combines the world of finance and business with environmentally friendly behaviour. Green finance is an arena for many participants, including individual and business consumers, producers, investors and financial lenders [1].

In recent years, China has adopted market-based means, such as green finance, as important means of improving the environment and strives to create a green credit policy system to support the development of a green economy, circular economy and low-carbon economy. Significant results have been achieved in promoting economic restructuring and industrial structural transformation and upgrading.

J. Liu
College of Finance and Trade, Guangdong Industry Polytechnic, Guangzhou, Guangdong, China
e-mail: gzliujunyan@126.com

J. Fu (✉)
College of Economics, Institute of Resource, Environment and Sustainable Development, Jinan University, Guangzhou, China
e-mail: fuan2@163.com

J. Fu and A. W. Ng (eds.), *Sustainable Energy and Green Finance for a Low-carbon Economy*, https://doi.org/10.1007/978-3-030-35411-4_8

According to statistics of the China Banking Regulatory Commission (CBRC), the green credit of 21 major banks and financial institutions of China reached 8.22 trillion Yuan by the end of June 2017. According to the Chinese green credit statistics system, green credit consists of two parts: one part supports manufacturing terminals of three strategic emerging industries, such as energy conservation, environmental protection and new energy vehicles; the other part supports projects and services of energy conservation and environmental protection, and this portion of loans is expected to save 215 million tons of standard coal annually and to reduce carbon dioxide emissions by 491 million tons annually.

As a means of green credit, the Guangdong Energy Efficiency Power Plant (EEPP) project is the first pilot project between the Chinese government and the Asian Development Bank (ADB) in the field of energy conservation, and it also known as the project of the ADB loan to promote energy conservation and emission reduction. To promote the national energy conservation and emission reduction efforts, the Chinese government began to discuss with ADB how to build the first EEPP project in China at the beginning of the 11th Five-Year Plan. In view of the important role of the Guangdong Province in China's economic development and energy consumption, and due to the pioneering spirit of being the first, Chinese government authorities and ADB agreed to consider Guangdong Province as a pilot province of the EEPP in March 2006. It was anticipated that Guangdong would explore financing and management models for energy conservation projects that can be replicated in other provinces and regions in order to promote energy conservation and emission reduction efforts throughout the country.

The context of the EEPP is to use ADB loans to carry out energy-saving and emission reduction efforts. The project, guaranteed by the Chinese government and using ADB's $100 million loan and supporting domestic funds, continues to support energy conservation and emission reduction projects in Guangdong Province on a rolling basis between 2009 and 2026. The project execution centre shall transfer the loan to each subproject implementing unit after the loan is exchanged into RMB.

Energy-saving efforts in Guangdong Province play an important role in the effectiveness of energy-saving efforts in China because Guangdong Province accounts for approximately one-tenth of the country's energy consumption. In short, the successful promotion of the EEPP has played an important role in promoting energy conservation and emission reduction in Guangdong Province and in the whole country.

2 Literature Review

2.1 Definition of Green Finance

Today, the term green finance appears in many studies. However, there is no universal definition of green finance. Höhne et al. [2] argue that green finance is a broad term that can refer to financial investments flowing into sustainable development projects and initiatives, environmental products and policies that encourage the development of a more sustainable economy. Zadek et al. [3] argue that green finance is often used interchangeably with green investment. However, in practice, green finance is a wider lens including more than investments, as defined by Bloomberg New Energy Finance and others. Lindenberg [4] suggests that green finance comprises financing of green investments, financing of public green policies and a green financial system. Volz et al. [5] argue that green finance comprises all forms of investment or lending that take into account environmental impact and enhance environmental sustainability.

In general, most studies agree that green finance is a kind of financial product or service for sustainable development projects and environmentally responsible investments [6]. Green finance is indeed a broad term that includes all forms of investment or lending related to sustainable development projects.

Green credit is frequently used in China and refers to the actual green finance product and service offered by banks in China, such as loans related to environmental protection, emission reduction and energy conservation projects. If green finance is considered as a solution to environmental problems and resource management [7, 8], then green credit is the actual bridge that connects the environmental industries with the financial institutions [9].

2.2 *Factors Affecting Green Finance*

The promotion of green credit as a quasi-regular tool is not new. Many nations, including developing countries, such as India and Brazil, use this tool to promote environmentally beneficial financing at subsidized rates, often using public financial agencies as conduits [10].

While capital cost is largely influenced by the macroeconomic environment and fixed costs are stable, the ability to minimize bad debt can be managed to increase the profit of a bank. Typically, this strategy requires the lenders to monitor a borrower's capital stock, earnings, liquidity, etc. Those factors are often recognized as counterparty credit risks that influence the borrower's ability to repay the loan (i.e. the default risk of a borrower) [11–13]. Recently, environmental risk is recognized as one of the most important factors that can influence credit risk.

As suggested by Thompson and Cowton [14], the consideration of environmental issues is mainly motivated by the concern of risk management. In addition, some scholars find a correlation between sustainability performance and credit risks of borrowers [15, 16]

There is increasing pressure being placed on firms from a number of different sources, such as governmental regulation, community participation and market demand, to engage firms in environmental initiatives. These factors play different roles at various developmental periods. Government regulation was initially the major pressure. However, community participation and market demand have become increasingly important [17].

The degree of participation in green finance is truly management's decision. According to Koch et al. [18], decision-making from a bank's perspective involves managing: (1) bank performance, (2) securities and interest rate risks, (3) liquidity and capital, (4) loans and credit and (5) investments, globalization and technology.

This article will analyse the factors affecting green finance by examining a case of green credit in Guangdong Province, China, namely, the EEPP.

3 Operations of the EEPP

3.1 Defining an EEPP

An efficiency power plan is a virtual power plant composed of specific measures in demand-side management that enables savings in electricity resources by adopting an investment on a package of energy-saving plans and energy efficiency projects. The reduced demand for electricity is treated as the amount of electricity provided by a "virtual power plant", and the international energy industry has implemented similar demand-side management of electricity. The ability to implement power demand-side management and dispatch the resources of the demand side is perceived as EEPP by the international energy community.

3.2 Introduction of the EEPP

In recent years, China has produced a series of green policies, including green tax, green procurement, and green policies that are relevant to the financial sector, namely green credit, insurance and security policies. Of the three, the green credit policy is the most advanced, with three agencies (the Ministry of Environmental Protection, the Peoples' Bank of China and the China Banking Regulatory Commission) sharing the responsibility for implementation [19].

As a means of green credit, the development of EEPP is an important measure to develop green credit and practise green development in China. The concept of EEPP was proposed in the opening year of China's 11th Five-Year Plan; it conforms to the requirements of energy-saving situations and will span the 12th Five-Year Plan and the 13th Five-Year Plan. The main goal of the project is to encourage the main body of energy conservation to carry out energy-saving and emission reduction activities and to support the development of energy-saving service companies. The project is consistent with the medium- and long-term energy conservation plans of Guangdong Province, the 12th Five-Year Plan for energy saving and longer-term development planning. The project will be one of the important examples of energy-saving efforts in Guangdong Province for a long time to come.

3.3 Organizational Structure and Function

The implementation, evaluation and supervision of EEPP are relatively independent, and the fund management is relatively separate from project management. The organizational management structure of the project is shown in Fig. 1, and the figure shows a clear division of labour among the departments and their respective duties.

The project coordination group is responsible for guiding the overall operation of the project, researching and formulating the project policy and reviewing the loan of the subproject. The project coordination team is composed of members from the Economic and Information Commission of Guangdong Province (EICGP), Guangdong Provincial Development and Reform Commission (GPDRC), Department of Finance of Guangdong Province (DOFGP) and State-owned Assets Supervision and Administration Commission of Guangdong Province (SASACGP). Among these members, the EICGP is the lead unit of the project and is responsible for the daily management and coordination of the project coordination group; the GPDRC is responsible for examining and approving the feasibility study report of the project, reviewing the application report of the project funds and reporting to the approval of the national Development and Reform Commission; the DOFGP is the representative unit of project creditor's rights and liabilities, responsible for withdrawing funds from ADB, managing special accounts for the project, signing trust contracts with trust financial service institutions and monitoring the operation of loan funds; the

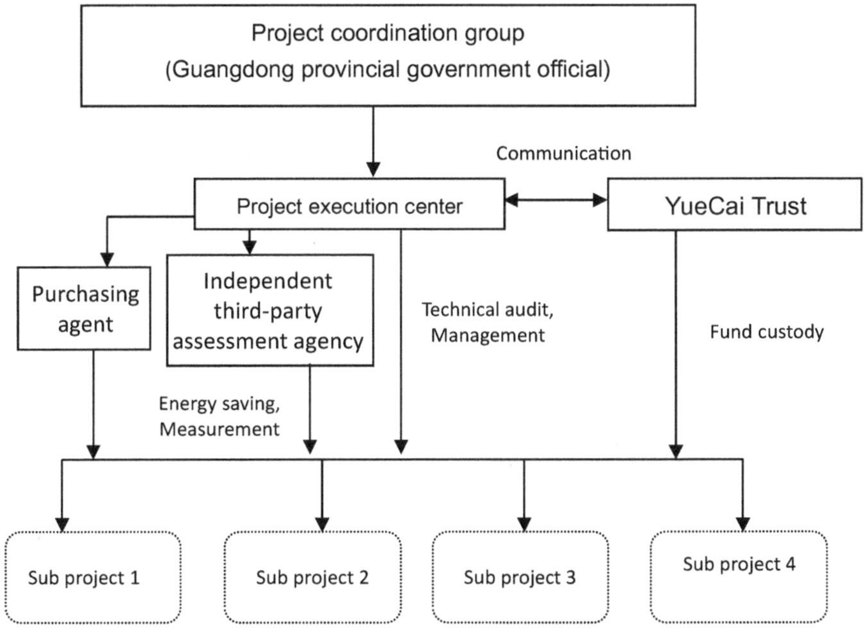

Fig. 1 Project organization management architecture. *Source* GEEPP [20]

SASACGP is responsible for project organization of the state-owned enterprises of the Guangdong Province.

The project execution centre (in the Energy-Saving Center of Guangdong Province) is in charge of project recruitment and concrete management. The specific work of this centre includes subproject collection, project overall monitoring and day-to-day project management, technical review and evaluation of subprojects, entrusting the third-party evaluation organization to evaluate the subprojects, submitting reports to the project coordination group and ADB according to regulations, etc.

The Guangdong YueCai Trust Co., Ltd. (GYCT) is the trust financial service institution. Entrusted by the Department of Finance of Guangdong Province, the GYCT is responsible for subproject unit financial capacity assessment, issuance and recovery management of subloans, portfolio management of subloans, mortgage or guarantee management of subloans, trust account management, submission of project financial reports on a regular basis, delivery of real-time information on the project and identification and reporting of problems in a timely manner.

Entrusted by the project execution centre, the third-party evaluation organization is responsible for the pre-project evaluation (estimation of energy savings, project implementation monitoring, etc.) and the post-retrofit evaluation (inspection and acceptance of the project, determination of the actual energy savings, follow-up tracking, etc.).

3.4 *Subproject Introduction of the EEPP*

Jointly promoted by the EICGP, GPDRC, DOFGP and SASACGP, the EEPP developed an overall operational framework for batch implementation, that is, the $100 million of total loans from the ADB was implemented in batches. At present, three batches of the first phase of the project have been completed, and $100 million of green loans have been used and repaid. The project has entered the second phase, that is, the batch of revolving use of loan funds. The subprojects for each batch are as follows.

3.4.1 The First Batch of the Project

The overall planning for Guangdong EEPP and its first batch of development were approved by the state council of China on 1 April 2008, and passed the resolution of the board of directors of ADB on 4 June 2008. The first batch (see Table 1) used $35 million in ADB loans, or 238 million Yuan (the agreed exchange rate is $1 for 6.8 Yuan), each for a loan period of three years. The ADB loans accounted for 68.1% of the total investment in subprojects, and all subprojects were completed in 2013. In the first batch, Guangzhou Zhiguang Electric Co., Ltd. and Guangdong Shaogang Songshan Co., Ltd. are the only large companies, and the remaining six companies are energy-saving service companies or small companies. Of the eight subprojects in

Table 1 First batch of the Guangdong EEPP

Subproject enterprise (borrower)	Content	Total investment (1000 Yuan)	ADB loan (1000 Yuan)	ADB loans account for total investment (%)
1	Promotion of high-voltage inverter	94,000	67,007	71.3
2	Building intelligent energy-saving reconstruction	25,000	20,000	80.0
3	Promotion of high efficiency and energy-saving transformer	12,500	10,000	80.0
4	Steam recovery and utilization/energy-saving retrofit of industrial boiler	12,500	10,000	80.0
5	Optimization monitoring and management system for transformer	51,600	20,000	38.8
6	Recovery and utilization of waste heat of flue gas in ring cooler	87,600	70,080	80.0
7	Promotion of reactive power compensation device	3741	2993	80.0
8	Promotion of the high efficiency and energy-saving transformer	62,500	37,920	60.7
Total		349,441	238,000	68.1

Source GEEPP [20]

the first batch, four are for the promotion and application of new equipment and new technologies, and four are for the upgrading of existing equipment and technologies.

3.4.2 The Second Batch of the Project

The fund application report of the second batch of the project was approved by the National Development and Reform Commission in November 2009. The second batch (see Table 1) used $22.06 million in ADB loans, or 150 million Yuan (the agreed exchange rate is $1 for 6.8 Yuan), each for a loan period of three years. The ADB loans accounted for 36.1% of the total investment in subprojects, and all subprojects were completed in 2014. In the second batch, Guangdong Real Faith Lighting Technology

Table 2 The second batch of the Guangdong EEPP

Subproject enterprise (borrower)	Content	Total investment (1000 Yuan)	ADB loan (1000 Yuan)	ADB loans account for total investment (%)
1	Promotion of solar photovoltaic power generation system	68,000	35,000	51.5
2	Promotion of energy-saving transformer	30,000	10,000	33.3
3	Promotion of regenerative high efficiency and energy-saving aluminium melting furnace	40,000	20,000	50.0
4	Digital substation transformation and distribution monitoring management system	127,800	20,000	15.6
5	Promotion of LED street lamp	100,000	50,000	50.0
6	Promotion of energy-saving automatic plastic moulding machine	50,000	15,000	30.0
Total		415,800	150,000	36.1

Source GEEPP [20]

Co., Ltd. and Borche Machinery Co., Ltd. are the only large companies, and the remaining four companies are energy-saving service companies or small companies. Of the six subprojects in the second batch, five were for the promotion and application of new equipment and new technologies, and one was for the upgrading of existing equipment and technologies (Table 2).

3.4.3 The Third Batch of the Project

The third batch (see Table 3) used the rest of the ADB loans, that is, $42.94 million or 292 million Yuan (the agreed exchange rate is $1 for 6.8 Yuan), each for a loan period of three to five years. The ADB loans accounted for 44.5% of the total investment in subprojects, and all subprojects were completed in 2015. In the third batch, Guangdong Shaogang Songshan Co., Ltd. and Borche Machinery co., Ltd. are the only large companies, and thc rcmaining four companies are energy-saving service companies or small companies. Moreover, this is the second time that these two large companies have participated in the Guangdong Energy Efficiency Power Plant project, which reflects the enthusiasm of participating in the project and demonstrates the appeal of the project. Of the six subprojects in the third batch, four were for the promotion and application of new equipment and new technologies, and four were for the upgrading of existing equipment and technologies.

Table 3 Third batch of the Guangdong EEPP

Subproject enterprise (borrower)	Content	Total investment (1000 Yuan)	ADB loan (1000 Yuan)	ADB loans account for total investment (%)
1	Promotion of steam recycling technology	94,813	64,000	67.5
2	Promotion of insulated thin-walled copper busbar	45,000	30,000	66.7
3	Energy-saving transformation of air conditioner	323,100	100,000	58.8
4	Promotion of high-power frequency conversion speed regulation system	75,000	50,000	66.7
5	Promotion of energy-saving automatic plastic moulding machine	102,000	37,000	36.3
6	Promotion of LED street lamp	16,000	11,000	68.75
Total		655,913	292,000	44.5

Source GEEPP [20]

3.4.4 The Batch of Revolving Fund of the Project

The first three batches were completed by 2015, and the ADB loan of US $100 million was used once; then, the project entered the batch of revolving fund (see Table 4). The batch began in August 2010 with the first subproject to repay the first principal amount and will continue until the end of the project in 2026. The number of subjects of the batch reached 22 by 31 October 2017, and rolling use of loans in Guangdong Province can promote more energy-saving and emission reduction projects. This batch was Guangzhou Zhiguang Energy-Saving Company's and others' second time participating in the project, Borche Machinery Co., Ltd.'s third time participating in the project and Zhuhai Industrial Green Building Technology Company's and others' fourth time participating in the project. The multiple participation of enterprises fully reflects the appeal of the EEPP to enterprises and the promotion of energy saving and emission reduction in Guangdong. The batch of revolving fund had used more than $100 million by 31 October 2017, meaning the ADB loans had been recycled twice. This observation reflects the project operation mechanism of recycling and continuously promoting energy-saving and emission reduction work in Guangdong.

4 Results and Findings

Different from traditional financial projects, the project of the Guangdong EEPP has distinct green financial characteristics from its origin of operations. The details are as follows:

4.1 The Purpose of the Project as Green Development

As a green finance project, the purpose of green development is to distinguish the EEPP from the traditional financial project. The main purpose of traditional financial projects is profit, while green financial projects are green environmental protection; there is an essential difference between the two goals. The three main application conditions of the EEPP are: a full mortgage guarantee, content of the project that conforms to requirements and measurement of energy savings. From the application conditions, we can see that the EEPP is a typical green project that aims to save energy and reduce emissions.

The technical scope and environmental protection requirements of the EEPP reflect green environmental protection. The EEPP has expanded the concept of the energy efficiency, aimed not only at saving electricity but also to the broad sense of energy conservation through the adoption of various energy efficiency and new energy development technologies to achieve the purpose of energy saving and emission reduction. The technical scope includes the optimal control of motor and motor drive systems and electric power transmission and distribution, such as transformer

Table 4 Revolving fund of the Guangdong EEPP

Subproject enterprise (borrower)	Content	Total investment (1000 Yuan)	ADB loan (1000 Yuan)	ADB loans account for total investment (%)
1	Promotion of energy-saving micro motor	50,000	30,000	60.0
2	Promotion of frequency conversion speed regulation system	45,000	30,000	66.7
3	Promotion of energy-saving transformer	20,000	12,000	60.0
4	Promotion of energy-saving industrial boiler	15,000	10,000	66.7
5	Construction of solar photovoltaic power station	28,818	15,000	52.1
6	Transformation of environmental degradation product line	27,350	20,000	73.1
7	Promotion of energy-saving dyeing equipment	50,000	20,000	40.0
8	Guangdong Yuedian Hongfa Hydropower Project	147,920	60,000	40.6
9	Construction of solar photovoltaic system and application of new energy in Guangdong province	244,000	60,000	24.6
10	Renovation of high efficiency LED lamps	66,240	40,000	60.4

(continued)

Table 4 (continued)

Subproject enterprise (borrower)	Content	Total investment (1000 Yuan)	ADB loan (1000 Yuan)	ADB loans account for total investment (%)
11	Renovation of intelligent control of central air conditioner	15,460	10,000	64.7
12	Renovation of high efficiency LED lamps	77,050	50,000	64.9
13	Guangdong Yuedian Dianbai wind power project	487,500	50,000	10.3
14	Energy-saving project of building system	227,086	20,000	8.8
15	Promotion of energy-saving injection moulding machine	30,000	20,000	66.7
16	Energy-saving technology and equipment for ceramic powder	28,164	15,000	53.3
17	Technical transformation project of recycled plastics	150,000	95,000	63.3
18	Promotion of solar photovoltaic power generation	74,814	50,000	66.8
19	Construction of solar photovoltaic power generation system and development of new energy	350,000	60,000	17.1

(continued)

Table 4 (continued)

Subproject enterprise (borrower)	Content	Total investment (1000 Yuan)	ADB loan (1000 Yuan)	ADB loans account for total investment (%)
20	Promotion of energy-saving distribution transformer	43,000	30,000	69.8
21	Promotion of solar photovoltaic power generation	50,400	35,000	69.4
22	Promotion of energy-saving injection moulding Intelligent equipment	40,000	19,000	47.5
Total		2,267,802	751,000	33.1

Source GEEPP [20]

and reactive power compensation, green lighting, heating, ventilation and air conditioning (HVAC) systems and other energy system optimization, air compression system and pump system energy saving, industrial waste energy recovery and utilization, industrial boilers and heat and power co-supply, etc. The three main environmental restrictions of the EEPP are as follows: not involving any land expropriation or involuntary immigration or causing any adverse effects on local residents; not being located in any particular environmentally protected area; and being in compliance with the national and provincial social and environmental protection-related laws and regulations design and implementation. From the technical scope and environmental protection requirements of the EEPP, it can be seen that the project content of the EEPP also aims to save energy and reduce emissions, which embodies the concept of green development.

4.2 Achieving Green Development

As a green financial project, promoting the harmonious development of the economy and environment and embodying the concept of green development are the essential characteristics of the Guangdong Energy Efficiency Power Plant project, which is different from those of traditional financial projects.

4.2.1 Reducing Energy Consumption in Guangdong Province

The energy-saving capacity already achieved by the project far exceeds the requirements of the project. During the pre-evaluation phase of the project, the project team signed a financing framework agreement with ADB to confirm the project energy-saving requirements (see Table 5). After the implementation of the project, the project team commissioned a third-party evaluation organization to evaluate the energy savings of completed projects according to Environmental Policy (2002 edition), Environmental Assessment Guidelines (2003 edition) and the domestic requirements of Methods and Parameters of Economic Evaluation of Construction Projects (third edition). By the end of 31 October 2017, the annual energy-saving capacity of the completed projects exceeded the ADB requirements (see Table 5). If we assume that the EEPP works 330 days a year and 24 h a day and use the 2016 national average industrial electricity price (0.7 Yuan per hour), then the EEPP could save end-users $355 million a year on electricity. Since the ADB loan can be recycled at least three times in the Guangdong Province within 15 years, the energy-saving benefit will be doubled, and the remarkable energy-saving benefit will promote the decrease of energy consumption in Guangdong. Thus, the project will bring good economic benefits.

The successful operation of the EEPP has greatly promoted the development of Guangdong's green economy. The EEPP was proposed at the beginning of the 11th Five-Year Plan, which conforms to the requirements of the current energy-saving situation and will span the 12th and 13th Five-Year Plans. The project is consistent with the energy conservation plan of Guangdong Province in the medium and long term in alignment with the energy-saving plan of the 12th Five-Year Plan and its long-term development direction. The EEPP will still be one of the important examples of energy conservation work in Guangdong Province for a long time to come. In the face of the difficult task of energy saving and consumption reduction, Guangdong Province has completed the energy-saving targets of the 11th Five-Year

Table 5 Energy-saving details of the EEPP

	Energy-saving requirements	Actual energy saving (by 31 October 2017)	Meet the requirements or not
Annual reduction in coal consumption (tce)	175,813	503,941	√
Annual reduction in electricity consumption (MWh)	532,767	1,527,093	√
Equal to EEPP (MW)	107	305	√
Annual electricity savings for end-users (million dollars)	42.6	355	√

Source GEPPP [20]

Plan and the 12th Five-Year Plan with the efforts of all parties concerned, and the energy consumption level is in the leading position in the country. The EEPP plays a significant role.

4.2.2 Economic Benefits Through the EEPP

First, industrial and residential electricity prices are not the same. Second, industrial electricity is different from peak value, average value and lowest value, and there is a ladder price for residential electricity. Finally, electricity prices are different in all regions of the country and in Guangdong. To simplify the calculation, this paper selects the average value of the national industrial electricity price in 2016, which is approximately 0.7 Yuan/KWh.

It is assumed that the depreciation period of the fixed assets is 10 years. According to the sixtieth provision of the Enterprise Income Tax Law of the People's Republic of China (2011 edition), the minimum age of depreciation for fixed assets of machinery, machinery and other production equipment is 10 years, so this paper sets the fixed assets depreciation life as 10 years.

The interest rate is chosen as the one-year deposit rate of the central bank in 2017. Because the fixed asset depreciation life is set to 10 years, we need to convert the project income into the current value of the current year. The interest rate used to calculate the present value is chosen as the one-year deposit interest rate of the central bank in 2017, that is 1.5%.

Then, the formula for calculating the total return on investment is:

$$\text{Total return} = \sum_{i=1}^{10} \left(\text{Annual electricity savings} \times \text{Electricity price} \div (1 + 1.5\%)^{\mathrm{i}-1}\right) \quad (1)$$

Second, we must calculate the total cost of the investment. Total project investment includes self-financing and ADB loans; enterprise self-financing does not calculate interest. Based on the actual cost of using the ADB loan, assume that the interest on the ADB loan is 2%. Then, the formula for calculating the total cost of investment is:

$$\text{Total investment cost} = \text{Enterprises own capital} + \text{ADB loan} \times (1 + 2\%) \quad (2)$$

Finally, we must compare the total return of investment and the total cost of investment. Based on formula (1) and formula (2), the investment returns of the first three batches of EEPP can be calculated (see Table 6). As seen from the table, 18 of the first three batches of 20 subprojects are profitable, with a profit rate of 90, and only two subprojects generate losses, with a loss rate of 10. The first three batches of 20 subprojects are profitable as a whole, and the proportion of income investment is 4.0; the EEPP has achieved good economic benefits.

Table 6 Investment returns of the first three batches

Batch	Subproject enterprise (borrower)	Total return (1000 Yuan)	Total cost (1000 Yuan)	Return/Cost	Profitable or not
1	1	445,233	95,340	4.7	√
	2	82,527	25,400	3.2	√
	3	15,706	12,700	1.2	√
	4	43,940	12,700	3.5	√
	5	144,532	52,000	2.8	√
	6	830,643	89,002	9.3	√
	7	413,074	3801	108.7	√
	8	83,471	63,258	1.3	√
2	1	20,299	68,700	0.3	×
	2	40,854	30,200	1.4	√
	3	95,068	40,400	2.4	√
	4	261,485	128,200	2.0	√
	5	44,392	101,000	0.4	×
	6	110,571	50,300	2.2	√
3	1	479,319	96,093	5.0	√
	2	106,941	45,600	2.3	√
	3	2,067,296	325,100	6.4	√
	4	165,857	76,000	2.2	√
	5	223,354	102,740	2.2	√
	6	18,058	16,220	1.1	√
Total		5,692,621	1,434,754	4.0	√

Source GEPPP [20]

4.2.3 The EEPP Achieved Synergistic Reduction of Greenhouse Gas, Air Pollution and Suspended Particulates and Promoted the Development of a Green Environment

Facing the grim situation of air pollution, suspended particulate pollution and international tendency of independent greenhouse gas emission reduction, "cooperative control" of greenhouse gases, air pollution and suspended particulate has become the key policy choice for the world, especially developing countries that are in a period of industrialization, to address domestic pollution prevention and the greenhouse gas emission reduction obligations of international control.

Air pollution, greenhouse gases and suspended particulate emissions, mainly caused by burning fossil fuels, have the same root synchronization. The EEPP not

Table 7 Comparison of reduction capability of the EEPP

	Project reduction requirements	Actual emission (by the end of 31 October 2017)	Complete or not
Suspended particulate (ton/year)	1785	5345	√
Sulphur dioxide (ton/year)	4795	13,744	√
Nitrogen oxide (ton/year)	1066	3054	√
Carbon dioxide (ton/year)	415,560	1,191,133	√

Source GEPPP [20]

only promotes the development of new equipment and new technology and transformation of the old but also facilitates the reduction of greenhouse gases, air pollution and suspended particulate at the same time.

The project's capacity to reduce emissions is far beyond the requirements of the project. The EEPP requires not only the content of the project to meet the requirements of the energy efficiency power plant but also the amount of emission reduction accountable. In the evaluation stage, the project team and the ADB signed a "financing framework agreement", confirming the project energy requirements (see Table 5). After the implementation of the project, the project team entrusted a third party to quantitatively evaluate the project based on the "environmental policy" (2002 edition) and "Environmental Assessment Guide" (2003 edition), combined with requirements for emission reduction projects of the domestic Chinese "construction project economic evaluation methods and parameters" (third edition). By the end of 31 October 2017, the annual emission reduction capacity of the completed project exceeded the ADB requirements (see Table 7) and achieved a good effect on the improvement of the environment. Because ADB loans can be recycled in Guangdong at least three times within 15 years, the benefits and environmental will be doubled.

The EEPP has successfully created a synergistic reduction of greenhouse gases, air pollution and suspended particulates, which is good for environmental pollution control and land surface temperature control, and has strongly promoted the green development of Guangdong. The EEPP aims to reduce greenhouse gas emissions and carbon emissions by 2030 and to promote the Chinese peak and peak ahead of Guangdong Province; the Guangdong Energy Efficiency Power Plant project also aims to reduce air pollution and suspended particulate matter and to promote the air pollution control and environmental management in Guangdong.

To summarize, the EEPP has promoted the coordinated development of the economy and environment and embodies the concept of green development.

4.3 Alignment with Public Policy Orientation

As a green financial project, the policy orientation to the service entity economy is the third characteristic of the EEPP that differs from traditional finance. Profit is the main purpose of traditional financial; funds may flow to the real economy as well as to a fictitious economy, which is different from green finance. The report of the 19th National Congress of CPC noted that finance is the core of modern economy; we must deepen the reform of the financial system, enhance the ability of financial services in the real economy, improve the proportion of direct financing and promote the healthy development of a multi-level capital market. The EPP is an important experience and reform of green finance development. All 42 items (by the end of 31 October 2017, including the first batch of three projects and recycling projects) are concerned with upgrading and popularizing the application of new technology and equipment, the funds all flow to the real economy, and the loan amount is 1 billion 431 million Yuan, which accounts for 39% of the total investment.

4.4 Promoting Industrial Upgrading

As a green financial project, the clear use of funds is the fourth characteristic of the EPP. While a traditional financial project has no clear direction of investment, encouraging and guiding the development of the main energy-saving project and supporting the development of energy service companies are the main goals of the development of the EEPP. The use of funds has a clear property of green finance and support of industrial development. There are 20 projects in the first batch, including 7 original equipment and technology and upgrading projects and 13 new equipment and technology promotion and application projects, such as a high efficiency, energy-saving transformer and aluminium melting furnace.

The popularization and application of these new devices and new technologies not only save energy and reduce emissions but also improve the production efficiency of enterprises and upgrade the industry. Additionally, the application of a solar photovoltaic power generation system not only achieved energy savings and emission reductions but also completely changed the production and industrial structure of the electric power enterprises. The replacement of thermal power and hydropower has obvious characteristics of upgrading the industry and industrial chain.

The upgrading of the original equipment and the original technology has also promoted the upgrading of the industry. There are seven equipment and technology upgrade projects of the EEPP, such as steam recycling and recovery and the utilization of flue gas waste heat. These projects can save energy and reduce emission reductions, as well as cope with waste and turn it into an advantage, which will increase product value and profits and upgrade the industry.

4.5 Social Benefits

The project brought good social benefits to the participating enterprises. First, the enterprise had to adapt to the global green trend in response to national policies of energy-saving and reduction of emissions in the development of new energy. Participating in energy efficiency projects will reduce the level of enterprise energy consumption and environmental pollution, assume more social responsibility and improve the enterprise's reputation. These practices will not only enhance the company's public image but also increase the opportunities for the enterprise to receive more government fiscal support, incentives and subsidies of programme applications. Second, the participating enterprises became familiar with the loan procedures and operation practices of international financial organizations, such as ADB, through the process of this programme, which improved the management level of the enterprises. Finally, enterprises will improve their credit and further broaden financing channels to attract more support of domestic financial institutions through participation and completion of the government in cooperation with ADB, which is especially important for small and medium enterprises.

The project has good social benefits for the local people. This project takes the mode of multi batch cycle lending, and each subproject unit is located in ten cities in Guangdong Province. So, the implementation of the project brings a positive change to the location of the project, such as the LED street extension and building intelligent node transformation programme, which benefit the local people. The projects improved the ecological environment and promoted ecological consciousness and the local people's awareness of energy conservation and the environment. Local people's ecological behaviour will be encouraged through the implementation of the project, which lays a solid foundation for the ecological development of these areas into a virtuous cycle.

The project has achieved great economic and social benefits and has been highly affirmed and praised by the Chinese government and the ADB. In 2011, the ADB, in its official publication of the "energy efficiency intervention practice report", noted that the EEPP is the only project that made significant progress of all the ADB loan projects and the only project that measured and certified energy efficiency during its implementation. In 2013, the first batch of the project was awarded "highly satisfied project" among all completed projects, and the second batch of projects obtained the "2012 best performance Loan Project" award among all under-construction projects, which are the only energy efficiency projects to receive the privilege in the implementation of the Chinese ADB project, and the third batch of the projects won "the 2014 best performance award loan project".

4.6 Development of a Green Finance Industry Chain

The green financial industry chain includes financial capital, a project execution centre, funds and financial managing institutions, independent third-party evaluation of enterprise, energy service companies and project target enterprises (see Fig. 1). The success of the Guangdong EEPP as a green financial operation sets an example to promote the development of the green finance industry chain, and specific examples of performance are as follows:

4.6.1 Development of Energy-Saving Service Industry

The energy services industry is the supporting industry for enterprises and projects to provide services in energy savings and emission reductions with contract energy management. Generally, this industry provides a diagnosis and financing and transformation services for energy savings of energy-using units to obtain reasonable profits and investment recovery through energy conservation benefits sharing. Therefore, this industry is the starting point to achieve the goal of energy saving and emission reduction for the energy industry, but it is not well developed due to various reasons.

Most of the domestic energy-saving service companies are unable to provide a full mortgage guarantee because of light assets, which create difficulty in obtaining the support of green credit. Through financial innovation and other measures, the Guangdong EEPP projects support a number of energy service companies with new energy-saving technology in a new field with high credit. Six of 20 projects in the first three batches are projects of energy services companies; they received 220 million Yuan of the ADB loan, accounting for 32.4% of the total loans. Through the process of supporting these energy service companies in terms of low-carbon technology and energy savings, the Guangdong EEPP projects promote the development of the energy services market and energy service companies and therefore the overall development of the green financial industry.

4.6.2 Development of Third-Party Evaluation Institutions

Without the third party of authority, evaluation institution is an important factor to restrict the development of green financial projects. At present, audit institutions of energy saving are mainly constructed by energy service companies, accounting firms, industry associations and research institutions. Lacking third-party assessment agencies with a strong professional credibility, energy service companies and energy units will have differences in energy saving and efficiency calculations in the process of project operation. An energy-using company will also be intentionally default in green financial projects, affecting the completion of the whole green financial project. According to a survey of green financial default projects, 38.89% exist because of energy-saving effects, 94.44% of which are caused by the unacceptable measurement standards and methods used by the energy service company; 74.14% of the banks that operated with green financial projects admit the lack of an evaluation system, which restricted the credibility of enterprise financing.

From 2009 to 2026, Guangdong EEPP projects supported the green credit project with a loan of one hundred million Yuan from the ADB and participated with third-party evaluation agencies in each project. The Guangdong EEPP supports and promotes the development of a third evaluation party to promote the development of the green financial industry.

4.6.3 Green Development of Small- and Medium-Sized Enterprises or the Guangdong Model of Green Finance

The Guangdong EEPP project not only pays attention to large energy consumption but also focuses on the green development of small- and medium-sized enterprises and creates the Guangdong model of green finance. The Guangdong EEPP project has both large electric companies, such as Shaogang and Zhiguang, and a large number of small- and medium-sized enterprises and energy service companies. There are four projects that listed companies in the first three batches of 20 projects, and two projects are with large companies. There are 14 projects that belong to small- and medium-sized enterprises and energy service companies, and the loans from ADB are 366 million Yuan, accounting for 53.8% of the total loans, which fully demonstrates support for small- and medium-sized enterprises and energy service companies.

Small- and medium-sized enterprises were keys to promoting Chinese economic reform for approximately the past thirty years, which made great contributions to economic development in China. As the pioneer of Chinese reform, the number of small- and medium-sized enterprises of Guangdong Province in 2009 reached 813.4 thousand, becoming the pivotal position and social economic development in Guangdong's reform and ranking second place in China. Therefore, the goal of the Guangdong EEPP is to innovatively activate small- and medium-sized enterprises, which is a breakthrough as compared to the traditional international financial loans.

The Guangdong EEPP is a positive innovation in policy research, mechanism design and project management based on the actual situation of Guangdong. The EEPP supports small and medium enterprises to become main energy-saving and emission reduction parties that strongly promote Guangdong's energy-saving emission reduction efforts and achieve good economic and social benefits. Distinctive characteristics of energy saving and emission reduction in the project provide a good demonstration in the implementation of ADB in other provinces and a reproducible, trustworthy and successful experience for construction of efficiency power plant projects in other domestic provinces and the international community. The successful implementation of the project marks a breakthrough in the cooperation between the Chinese government and ADB in the field of energy saving and emission reduction. This project fully demonstrates that the Guangdong mode of SMEs is the model of energy saving and emission reduction and is fully consistent with China's national conditions.

5 Conclusions

Through the case study of the EEPP, we find that policy support is necessary in the early stage of green financial development.

As a green financial project, requiring policy support in the early stage of development is the sixth characteristic of the EEPP. The non-profit nature of green financial has an obvious disadvantage in market competition, especially in situations with bound budgets and weak public consciousness of green development compared with traditional financial; therefore, the development of green finance requires not only development of public awareness of environment protection but also the engagement of government support in development. The policy support of the EEPP includes mortgage innovation, energy reward, low financing costs, long loan period, government bared loans, external risk and project operation costs, so these policies provided by the EEPP are the main differentiations from the traditional financial project.

5.1 *Mortgage Guarantee Innovation*

Through the innovation of mortgage guarantee, the EEPP provides the necessary policy support to the development of green finance. Because the EEPP is national sovereign guaranteed, it should be strictly risk controlled and fully secured. Constrained by funding strength, the majority of small- and medium-sized enterprises and energy service companies will find it difficult to achieve full collateral requirements, while the energy conservation centre has collateral innovation, such as land guarantee, group guarantee, bank guarantee, pledge of accounts receivable (such as SPDB), executives of credit guarantee, executives real estate guarantee, joint guarantee (such as Guangdong 50 million loans, signed by third-party guarantees, a finance company fall back of repayment after 3 years), third-party guarantee and accounts receivable guarantee, to help the application of the EEPP for some small- and medium-sized enterprises and energy service companies.

5.2 *Energy- Saving Reward Funds*

The energy-saving incentive fund is another policy support provided by the EEPP for the development of green finance. Guangdong Province issued the "Guangdong Province to promote energy-saving emission reduction of ADB loan (EPP pilot) project management manual" to reward the subproject borrower who achieved energy savings and debt service on time. Generally, project companies receive incentive funds, such as 15 projects in the first 20 projects of three batches, and there were only five remaining projects with no reward funds for failing to complete the energy-saving targets or execution.

The reward funds are derived from the project surplus. Seventy per cent of project earnings are used to reward the project unit, and project earnings include loan interest, financial income and other income after deducting the project operation cost; the difference in loan interest comes from the differences between the interest rate of project group obtained loans from the ADB's (the first and second batch of loans' interest is LIBOR up 0.2%, the third is up to 0.3%. The average interest rate of the projects is approximately 1%) and the interest rate project offered to project enterprises (interest rate of 6 months' commercial loans issued by the people's Bank of China falls 10%); financial income is derived from a portion of funds that has not been loaned out by the fund custodian (Guangdong finance trust) in managed funds during the financial implementation.

5.3 Low-Cost Financing

The low cost of financing is the third necessary policy support provided by the EEPP to the development of green finance. The nominal interest rate project provided to subproject enterprises is 10% lower than the commercial loans interest rate provided by People's Bank of China, and the top three batches' project loan rate is only approximately 3% because of tax preferential policies (the implementation period is from 2009 to 2015 and adjusted according to the central bank interest rate); considering the energy reward, the real interest rate of the enterprises involved in the project is very low, generally less than 2%. For example, the interest is 3.8 million with a 67 million loan and 3.2 million reward for the final return of the project for the Guangdong power group. Therefore, the actual interest from the ADB loan is less than 1%, and some are even as low as 0.5%, far below commercial lending rates, which greatly reduce the cost of corporate financing and strongly support the development of green financial projects.

5.4 Long-Term Loans

Arrangement of long-term loans is the fourth necessary policy support provided by the EEPP for the development of green finance. Generally, the operation cycle of green financial projects is longer, which becomes an important feature and key factor that limits the development of green finance and differs from traditional financial projects. Funds are required in the pre-period of contract and investment in green finance projects, which will be used for purchasing equipment and construction, while loans for traditional financial projects will be offered after the first payment during the project operation. Green finance investment funds for projects cannot be effectively satisfied because of the mismatch of time, which often leads to the abortion of the project. The funds of the EEPP lend early with a period of 3–5 years, as well as a grace period; a three-year loan has a one-year grace period and the ability to pay the principal in four stages in the last two years with monthly interest; a five-year loan has a two-year grace period and a schedule to pay back the principal twice a year. The longer loan period and relaxed repayment arrangement reduce the financial pressure of the enterprises, thus bringing about economic benefits to the participating enterprises while facilitating successful operations of the project.

5.5 Government Undertaking the Risk

The government takes the external risk and operation cost of loans as the fifth necessary policy support for the development of green finance in the EEPP. The direct support of the government is also an important feature of the green financial project, which is different from traditional financial projects. The EEPP loans are in US dollars, while loans to participating enterprises in the project group are in RMB, whereas interest risk and exchange rate risk are borne by government alone but not by the enterprises. At the same time, the government set up a project coordination group responsible for the overall operation of the project and guidance of the formulation of project policy. The project execution centre is managed by the project coordination group. The Guangdong finance trust and third independent party organization are responsible for the project operation and commission. Without the project cost, including the operation costs and service charges of hiring intermediate financial services institutions, the third-party evaluation agency cost borne by the government can be reduced.

References

1. Yao W, Qiang Z (2016) The role of green finance in environmental protection: two aspects of market mechanism and policies. Energy Procedia 104:311–316
2. Höhne N, Khosla S, Fekete H, Gilbert A (2012) Mapping of green finance delivered by IDFC members in 2011. Ecofys, Cologne
3. Zadek Simon, Flynn Cassie (2013) South-originating green finance: exploring the potential. Geneva International Finance Dialogues, Geneva
4. Lindenberg N (2014) Definition of green finance. DIE mimeo. Retrieved from: SSRN: https://ssrn.com/abstract=2446496
5. Volz U, Böhnke J, Eidt V, Knierim L, Richert K, Roeber GM (2015) Empirical analysis of supply of and demand for green finance in Indonesia. Financing the green transformation. Palgrave Macmillan, London, pp 56–94
6. Cui Y (2017) Analyzing green finance incentives: an empirical study of the Chinese banking sector. Master's Thesis, University of Waterloo, Waterloo, ON, Canada
7. Li X, Xia G (2014) *Zhong guo lv se xin rong bao gao*. [China green finance report]. China Finance Press, Beijing
8. Wang Y (2013) *Lv se xin dai yan jiu yu fa zhan zong shu*. Research in green credit policy and development summary. Times Financ 7:224–225
9. Salazar JE (1998) Environmental finance: linking two worlds. Paper presented at a workshop on financial innovations for biodiversity. 10th Global Biodiversity Forum, Bratislava, Slovakia, 1–3 May 1998, 15 pp

10. Taylor RP et al (2008) Financing energy efficiency: lessons from Brazil, China, India, and Beyond. The World Bank, Washington, D. C
11. Caouette JB, Altman EI, Narayanan P (1998) Managing credit risk: the next great financial challenge, vol 2. Wiley
12. Fitch TP (1997) Dictionary of banking terms. Barron's Educational Series Inc. Hauppauge, New York
13. Saunders A (1999) Credit risk measurement: new approaches to value at risk and other paradigms. Wiley, New York
14. Thompson C, Cowton P (2004) Bringing the environment into bank lending: implications for environmental reporting. Br Account Rev 36(2):197–218
15. Bauer R, Hann D (2010) Corporate environmental management and credit risk. ECCE Working Paper. University Maastricht, The European Centre for Corporate Engagement
16. Goss A, Roberts GS (2011) The impact of corporate social responsibility on the cost of bank loans. J Bank Finance 35:1794–1810
17. Zhang B, Bi J, Yuan Z, Ge J, Liu B, Bu M (2008) Why do firms engage in environmental management? An empirical study in China. J Clean Prod 16:1036–1045
18. Koch TW, MacDonald SS, Edwards V, Duran RE (2014) Bank management: a decision-making perspective. CENGAGE Learning Asia
19. Aizawa M, Yang Chaofei (2010) Green credit green stimulus green revolution? China's mobilization of banks for environmental cleanup. J Environ Dev 19(2):119–144
20. Guangdong Efficiency Power Plant Project (GEPPP) (2017) http://www.gdepp.cn. Last accessed 31 October 2017

Emerging Green Financing Issues

Impact of Climate Risk Factors on Valuations of China A-Share Market

Entela Benz-Saliasi

Abstract An increasing number of companies have embraced ESG, and an increasing number of investors have focused on the ESG impact behind their capital allocation decisions. To date, however, prior academic literature has not distinguished the useful environmental indicators in China A-share market. This study investigates the effect of environmental indicators on the valuation multiples. Using a normalized sample of 222 Chinese companies co-listed in China A-shares and MSCI China Index, it finds that most of the environmental indicators are statistically significant. Total GHG CO_2 intensity per sales and percentage of water recycled has significance in determining P/B value. This study also finds that the higher the government involvement, the higher the energy inefficiency and the GHG CO_2. These results speak to the significance of environmental factors in company valuation in China A-share market and also have implications for asset managers who have committed to the integration of environmental factors in their capital allocation decisions.

Keywords Climate risk · Climate change · Investments · Financial performance · Environmental factors · GHG · Water · Waste · Energy inefficiency

1 Introduction

Since the turn of the millennium, the world appears to have increased its commitment to sustainability and future growth. Investors have been increasingly focused on the ethical impact behind the investment while achieving outstanding returns. Meanwhile, global sustainable investment assets have expanded dramatically from $13.3 trillion in 2012 to $22.9 trillion at the start of 2016 according to Global Sustainable Investment Review, 2017 report. However, only US$500bn of US$23trn ESG asset is managed in Asia, and 90% of that is in Japan. Asia ex-Japan stands at a low level of 0.2% of the total ESG AUM, a level that clearly has room to grow.

E. Benz-Saliasi (✉)
Department of Finance, School of Business and Management, The Hong Kong University of Science and Technology, Kowloon, Hong Kong SAR, China
e-mail: entelabenz@ust.hk

J. Fu and A. W. Ng (eds.), *Sustainable Energy and Green Finance for a Low-carbon Economy*, https://doi.org/10.1007/978-3-030-35411-4_9

Closer to home, recognizing the intensity and the severe consequences of climate risk the Chinese government has laid a very ambitious plan. China's 13th Five-Year Plan targets a carbon intensity reduction of 18% and energy consumption cap of 5 billions tons of coal equivalent for 2020. By 2030, they target to have 20% renewable energy and reduce the carbon intensity by 60%. These targets are critical given the role of China. According to BP Statistical Review of World Energy, in 2017, China is the most significant contributor to GHG global emissions holding 27% of the total, followed by USA and India with 15% and 7%, respectively.[1] Currently, in China, one of the most critical restricting factors on economic development falls into the natural resource scarcity resulted from resource imbalance and industrial pollution which alarmed the government and public. Taking the water risk, in China alone, around 145 million people are exposed to flooding risk due to sea level rise. Even more acute is this issue in the Guangzhou area and surroundings due to the very high concentration of human and economic capital.[2] The societal and economic implications are unmeasurable.

With the growth of the Chinese economy and opening-up procedure in the Chinese market, domestic and foreign investors could no longer ignore environmental risk. With the inclusion of Chinese A-share[3] stocks into global equity benchmarks, asset owners must assess whether standard systematic approaches to mitigate risk and generate excess equity returns are equally valid in China. These environmental risks are underappreciated and have soon started to unfold. Significant spending on sustainable infrastructure and government incentives would be in need to meet emission reduction targets. These would present considerable investment risks and opportunities through tightening regulations, changing consumer preferences and disruptions caused by the advancement in technology.

Betting on better risk management and alpha generation, over the past decades, one of the most critical trends in portfolio management has been the incorporation of environmental, social and governance (ESG) data that integrate a measurement system to capture potential inequalities, social risk and long-term performance. ESG is more than just an ethical way of investing. ESG can enhance the portfolio by substantially reducing volatility, increasing Sharpe ratios and limiting drawdowns and have some merit as stand-alone alpha sources according to [1]. Using a comprehensive meta-study, they conclude that in 90% of the cases there is a nonnegative correlation between ESG and company financial performance. As for individual studies just to mention one or two, companies that perform high on environmental indicators achieve a cost of equity [2] and the cost of debt reduction [3]. Further, a 25% reduction on carbon emission per sales reduces the cost of equity, on average, by 0.4 basis points [4]. Oikonomou et al. [5] find a negative (positive) relation between systematic risk and a measure of aggregate strengths (concerns) for S&P 500 firms. They also find that community, employment and environmental concerns are significantly and positively related to systematic risk. And the list continues.

---

[1] BP Statistical Review of World Energy 2018.

[2] Deutsche Bank: Measuring Physical Climate Risk for Equity Portfolios, 2017.

[3] https://www.msci.com/china.

While there is a growing body of evidence on risk/return impact of environmental factors, the results tend to be either weak or confusing. This is because too much focus is given to the model or robustness check, with little understanding of the indicators to be used, the data collected and the method to aggregate to date from the providers. The same type of indicator is very often measured in different ways, resulting in a different meaning and conclusion (Gonenc et al. [6]. To avoid dubious results, we employ widely used and quantifiable indicators, rather than ratings (such as ESG score) that often include activities that have no financially material implication on company level.

The primary objective of this study is to answer the following questions: What works and what does not when it comes to physical climate risk, closely related to environmental factors investing in China? Is there statistical importance of specific environmental indicators toward company financial performance in China A-shares? Most of the indicators are industry-specific; hence what would happen if we normalize per sales and control for the industry as well?

We show that there is indeed a statistically significant relationship between physical climate risk capture by environmental score (*E*-score) and company financial valuations. Furthermore, the most striking result is that the most GHG polluting and most water-intensive companies are also the ones with the highest government ownership (% SOE). Last, we believe that most of these risks are currently not prices, giving rise to opportunities in certain sectors.

2 Data and Sample Description

MSCI China A-share index companies are all part of China A-share market. As such, they enjoy high market capitalization and many other intangible elements such as the accessibility to foreign markets and connections with the government among others.

The industry group is classified under the guidance of the Global Industry Classification Standard (GICS®). Our sample constitutes 221 Chinese companies co-listed in China A-shares, and MSCI China Index deemed to be the most liquid among the China A-shares. There are eleven industry sectors: (a) communication, (b) consumer discretionary, (c) consumer staples, (d) energy, (e) financials, (f) health care, (g) industrials, (h) materials, (i) real estate, (j) technology and (k) utilities. Table 8 presents the sectors and the number of companies in each sector. The largest concentration of companies is found in financials and industrials with 55 and 41 companies, respectively. The lowest is in communication with only four companies.

2.1 Company Valuation Data Description and Analysis

The valuation multiples we choose are a price-to-earning (P/E) and price-to-book P/B ratios. P/E ratio is an equity valuation multiple (see Table 1). It is calculated by

Table 1 Sectors and companies in each sector (there are 11 sectors and 222 companies overall). P/E and P/B ratios are normalized per each sector—data as of July 2018

Industry	Number of companies	Dividend yield %	P/E	P/B	Average ownership ratio (% SOE)
Communications	4	0.59	25.00	1.28	59.2
Consumer discretionary	25	1.97	15.76	2.18	21.8
Consumer staples	12	1.72	28.89	5.14	21.0
Energy	7	3.33	14.81	1.22	54.3
Financials	55	2.85	8.87	1.12	39.6
Health care	13	0.97	28.21	3.41	23.1
Industrials	41	1.65	16.50	1.69	45.8
Information technology	25	0.77	33.80	3.00	10.8
Materials	10	1.59	16.94	1.95	36.0
Real estate	17	3.15	9.50	1.46	31.9
Utilities	12	2.14	21.37	1.49	51.0

dividing the market price per share by earnings per share (annually). It facilitates comparison between different companies within the same industry. Higher P/E ratio is often associated with growth stocks that are developing faster than average, which attracts investors. Lower P/E ratio often attracts the investors interested in investing in value stocks as low P/E ratio represents that the stock is available at a cheap cost. The other valuation multiple we use is the *P/B* ratio. The P/B multiple offers an insightful perspective on how the market evaluates a company's assets, comparing to its earnings. This makes it particularly useful for valuing firms with significant financial assets.

P/E and P/B ratios include information about other factors such as earning turnarounds, growth prospects, proportion of debt, management efficiency and investor sentiments. We improve the measurement by normalizing the P/E and P/B ratios using the MSCI China Sector Average P/E (P/B) score based on MSCI Onshore China A Index as benchmarks.

$$\text{Sector Normalized P/E Score} = ((\text{Company P/E Score})/(\text{MSCI China Sector Average P/E Score})) \times 100$$
$$\text{Sector Normalized P/B Score} = ((\text{Company P/B Score})/(\text{MSCI China Sector Average P/B Score})) \times 100$$

We also collect and analyze *the % Div. Yid* per industry (average dividend payout ratio of that industry) as a signal of investors' preference on short-term cash inflow in the form of dividends or long-term stock price appreciation due to the growth funded by retained earnings. Anecdotal evidence indicates that the state companies in China are more inclined to operate in an environment with the absence of hard budget constraints which is replete with moral hazard.

Last but not least, our analysis takes account of average ownership ratio (percentage of state-owned % SOE) as the government ownership proportion in terms of equity in each company. Given the top-down ESG approach in China, we believe that that is an essential driver of environmental and social policies implemented on the company as well as on countrywide level. It is also a factor to be priced in company valuations given its importance in the decision-making process on the board level.

To illustrate valuation multiples across different sectors, we compare the normalized P/E ratios and P/B ratios. In particular, we see IT sector owns the highest industry average P/E while financials owns the lowest. In terms of P/B, consumer staples' sections ranked as the top while industrials is the lowest. Fifty-five companies have been included under the financials. On the contrary, only four companies have been listed under the communication sectors. Among all the companies we have compared, energy sector owns the highest average ownership ratio which aligns with our expectation. The following two graphs are based on the data we have collected (see Figs. 1 and 2).

Both figures point to an interesting result. For China A-shares included in the MSCI index, the relation of average ownership ratio is adverse to the company valuations captured by P/E and P/B ratios. Put it differently, the more independent the company, the higher the valuations. For example, information technology and consumer staples with the lowest ownership ratio (10–20%) enjoy the highest P/E valuations (~30). The opposite is true for energy, communications and industrials.

On the dividend yield ratio, from a shareholder perspective, both charts point to a positive correlation between average ownership and dividend yield. Dividends are largest in the companies with highest government ownership. One possible explanation is that these companies are beyond the growth stage. Hence, earnings are distributed rather than retained for growth.

Fig. 1 Cross-industry comparison, China A-shares market, LHS: P/E, Div. Yid%; RHS: average ownership ratio. In brackets are a number of companies per sector. Normalization as of July 2018

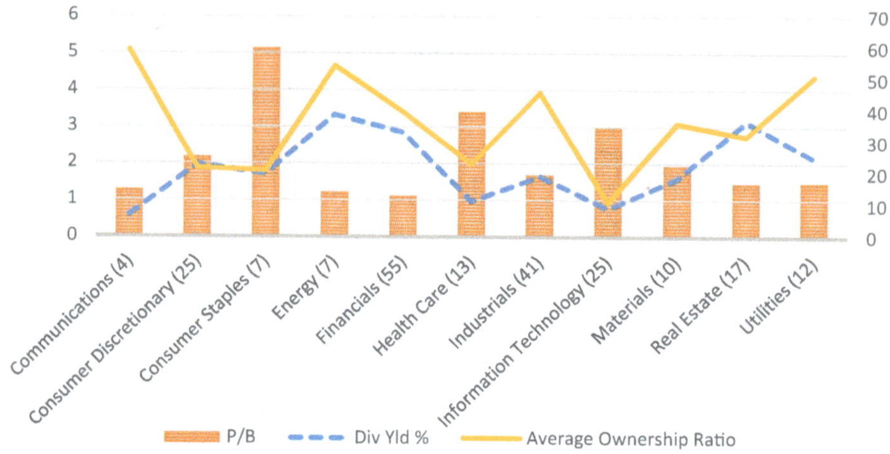

Fig. 2 Cross-industry comparison, China A-share market, LHS: P/B, Div. Yid%; RHS: average ownership ratio. In brackets are a number of companies per sector. Normalization as of July 2018

The following sections aim at analyzing the potential relationship between the environmental factors and company valuation as well as the role/influence, if any, of the government ownership ratio.

3 Company Environmental Data Description and Analysis

Bloomberg has been adopted as the data provider[4] for the environment indicators (*E*-scores). Based on BlackRock's model and the emphasis outlined by the Chinese government's Five-Year Plan, five indicators are chosen to quantify the environment risk:(i) CO_2 (total GHG CO_2 emission intensity per sales), (ii) *energy* (energy intensity per sales), (iii) *water* (water intensity per sales), (iv) *waste* (waste generated per sales) and (v) *recycled* (percentage of water recycled) (see Table 2). The first four indicators are divided by total sales which stands for a good proxy for efficiency rather than the absolute level.

Table 3 shows how these indicators behave across different sectors of MSCI China A-share companies. Sensitive industries perform a higher level climate-related risk.

i. None of the four companies in communication sectors reports the key environmental indicators.

[4]The platform is on track to provide company-reported ESG data for almost 9500 companies in 83 countries. It has been broadly used by over 700 reports and one million unique users per month, covering four ESG investment themes including 35 ESG indicators where eleven of them are E factors. The eleven indicators are aimed to measure the direct or indirect impact of a company's activity on the environment (waste management, level of carbon dioxide emissions, responsible consumption of water, development of renewable energy, degree of energy efficiency are examples).

Table 2 Environmental (climate) risk indicators used for the analysis. There are five environmental indicators (E_score_i) used, and we denote each as i

Environmental indicators (E-score$_i$)	Definition
1. CO_2	Total GHG CO_2 emission intensity per sales
2. Energy	Energy intensity per sales
3. Water	Water intensity per sales
4. Waste	Waste generated per sales
5. Recycled	Percentage of water recycled

Table 3 Cross-industry comparison of environmental indicators. CO_2, energy, water and waste are reported as total number per sales. For example, CO_2 is the total GHG emission divided by sales. The recycled indicator is in percentage only

Industry average E-scores	CO_2	Energy	Water	Waste	Recycled
Communications	-	-	-	-	-
Consumer discretionary	108.94	341.34	113.28	2.69	98.40
Consumer staples	26.59	38.19	-	5.50	-
Energy	70.71	203.19	-	212.48	-
Financials	0.32	10.64	5.23	-	-
Health care	4.12	40.56	391.15	2.23	-
Industrials	92.21	224.24	136.93	1.71	10.58
Materials	32.19	469.07	8595.39	1219.57	89.61
Real estate	-	3931.59	-	-	-
Technology	1.67	9.30	354.33	0.71	61.71
Utilities	-	4624.05	2749.91	0.04	-

ii. Consumer discretionary and industrials pollute the most in terms of CO_2 emissions. This result is consistent with the ones reported in [7]. Working on a comprehensive sample of 1600 firms, from 43 countries with data of 2008–2016, they also found that the industrials and consumer goods sectors are the highest CO_2-emitting industries.
iii. Utilities and real estate are the least efficient in terms of energy use, while materials and utilities are the least efficient in water usage. This is important in light of hefty fines and stricter regulation in China due to high water scarcity and water pollution.
iv. Materials are also the sector with the highest waste per given sales.
v. Overall, materials and utility sector have the highest environmental risk in China.

While some researchers argue that sensitive industries produce better ESG performance [8], it may not be the case in China. Since the overall disclosure is scarce, there is little incentive for the companies in sensitive industries to disclose and improve their environmental performance in order to maintain or improve reputation. This simple analysis has important consequences: Investors looking to put money in these sectors need to quantify, price and undertake stress scenarios and proper risk analysis to understand environmental risk in their portfolios better.

Getting more granular, within sectors, the company environmental performance, hence indicators (*E*-score) should be evaluated at a relative horizontal level, i.e., peer comparison. This is because the central added value of E within the ESG investing comes from how well the firms managed their industry-specific environmental risk while controlling for the differences of regulations, market landscape and operational risk level across industries. Consequently, based on the ESG data provided by Bloomberg, we normalize each environmental score using the following method:

$$\text{Sector Normalized } E\_\text{score}_i = \left(\frac{\text{Company } E\_\text{score}_i}{\text{Industry Average } E\_\text{score}_i}\right) \times 100$$

Thus, environmental leaders and laggards in sectors could be identified. Except for *% recycled*, the larger the level of the score, the higher the level of the environmental indicators used and therefore the higher the climate risk the company holds.

4 Results and Interpretations

In an attempt to explain the statistical importance of the above environmental (*E*-score) indicators on P/E and P/B ratios of the firms listed in China A-shares, we firstly ranked the stocks of the entities we have chosen in descending order according to a specific *E*-score. The lower the *E*-score, the better is the company managing its environmental risk and more specifically the physical part of climate[5] risk. Two groups are established by taking out the first quartile (top 25% in the specific *E*-score) and the fourth quartile (last 25% in the specific *E*-score). We then calculate the mean and the standard deviation for each quartile. Finally, we make use of the difference in mean methods to conclude whether there is a statistically significant difference between the two groups/quartiles.

To provide readers with explicit evidence of association on an absolute level, meaning no sector normalization, we conduct the hypothesis test on the P/E and P/B levels before normalizations (Table 9 in Appendix 1). The companies are ranked in descending order in terms of market capitalization. The hypothesis is conducted within the first quartile and the last quartile of the companies by market capitalization. Interestingly while the difference in the P/E ratios is statistically significant (but not on P/B), none of the *E*-scores indicates significant differences. This means that unlike the ESG ratings, the performance on our environmental indicators is not dependent on the company's market size. This is good news as it shows that normalizing the level of environmental indicators per given sales, it is neutralizing market capitalization.

[5]The climate risk we mention here is actually the physical risk as mentioned and measured in the previous paper.

4.1 Total GHG CO_2 Emission Intensity Per Sales

We start the analysis with total GHG CO_2 emission intensity per sales (CO_2) versus the P/E and P/B ratios. Given that some companies do not report any data on this specific E_score_i, our sample shrinks to 14 companies, 7 with high CO_2 versus 7 companies with low CO_2 emissions.

Our first hypothesis is as follows:

$$H_0 : \text{Difference in normalized } E\_\text{Score}(\text{CO}_2)\text{in two groups} = 0$$
$$H_1 : \text{Difference in normalized } E\_\text{Score}(\text{CO}_2)\text{in two groups} \neq 0$$

The t-statistic is evaluated by:

$$t\text{-statistics} = \frac{(x_0 - x_1)}{\sqrt{\frac{\text{SD}_0^2}{n} + \frac{\text{SD}_1^2}{n}}}$$

where n is equal to seven in this case and the resulting t-statistics on $E\_\text{Score}(\text{CO}_2)$ is 11.26 which is higher than 2.02 (for small sample size). Thus, the null hypothesis can be rejected at 90% significance level. It can be concluded that $E\_\text{score}_{\text{CO2}}$ is statistically different among the top 25% quantile and the last 25% quantile.

Similarly, our second hypothesis tests the significance of the mean difference of P/E and P/B ratios between top versus bottom companies. When comparing the companies ranked on the top 25% and the last 25% of the $E\_\text{Score}(\text{CO}_2)$, the difference of mean t-statistic on P/E ratio and P/B ratio is −1.38 and −2.05, respectively. Hence at 90% confidence interval, only valuation captured via P/B is statistically different. From valuation perspective, it indicates that polluting companies have lower price-to-book ratios.

Table 4 presents *t*-statistics and top/last quartile stocks in China A-share market ranked based on CO_2 intensity per sales as our chosen environmental indicator. We analyzed 222 China A-share companies part of the MSCI EM Index. The data are as of July 2018. The P/E and P/B ratios are normalized per sector. Div. Yid % is the sector average dividend yield. Government ownership is reported in the third column. We ranked the companies based on CO_2 per given sales normalized per industry level as well (CO_2 industry normalized). We then calculate the difference of means to find if there is a statistically significant difference between top versus bottom quartile companies in terms of the environmental indicator and financial variables presented in the table. We find that the difference of means is significant in terms of CO_2 industry normalized usage, ownership ratio and firm performance P/B ratio. It is nonetheless not significant for the P/E ratio. The significance of the difference of means is measured via t-statistic T0 at a confidence interval of 99%, 95% and 90% and represented with (***), (**) or (*), respectively.

Furthermore, there is a statistically significant difference in the ownership ratio between top versus bottom, implying that the polluters tend to be mostly state-owned

Table 4 *T*-statistics and top/last quartile stocks in China A-share market ranked based on CO_2 intensity per sales

CO_2 intensity per sales

Company	Industry	Ownership	Industry normalized PE	Industry normalized PB	CO_2 intensity	CO_2 industry normalized	Div. Yid %	P/E forward
COSCO Ship Dev A (HK-C)	Industrials	91.47	0.84	0.52	259.01	2.81	1.65	12.28
China Citic BK A (HK-C)	Financials	92.45	0.80	0.74	0.71	2.18	2.85	7.73
Air China A (HK-C)	Industrials	73.16	1.39	1.23	194.45	2.11	1.65	12.28
China East Air A (HK-C)	Industrials	85.67	1.13	1.32	190.33	2.06	1.65	12.28
GF Securities A (HK-C)	Financials	24.11	1.66	1.34	0.67	2.06	2.85	7.73
China Southern Airlines Co., Ltd	Consumer discretionary	82.25	1.26	1.10	197.80	1.82	1.97	12.56
China Gezhouba A (HK-C)	Industrials	84.82	0.49	0.89	156.66	1.70	1.65	12.28
	Mean	76.28	1.08	1.02	142.80	2.11	2.04	11.02
	SD	22.10	0.37	0.29	94.14	0.33	0.52	2.08
China Commu Con A (HK-C)	Industrials	5.60	0.71	0.68	13.32	0.14	1.65	12.28
China Intl Mar A (HK-C)	Industrials	16.37	1.71	1.33	12.93	0.14	1.65	12.28
Haitong Sec A (HK-C)	Financials	36.00	1.93	1.12	0.04	0.11	2.85	7.73

(continued)

Table 4 (continued)

CO_2 intensity per sales								
Company	Industry	Ownership	Industry normalized PE	Industry normalized PB	CO_2 intensity	CO_2 industry normalized	Div. Yid %	P/E forward
CRRC Corp A (HK-C)	Industrials	85.15	1.93	1.61	9.38	0.10	1.65	12.28
China Life Ins A (HK-C)	Financials	3.67	3.04	2.39	0.03	0.08	2.85	7.73
Huatai Securities Co., Ltd	Financials	13.33	1.50	1.26	0.02	0.06	2.85	7.73
CSSC Offshore and Marine Engineering Group	Industrials	90.70	32.06	2.015	5.37	0.06	1.65	12.28
	Mean	35.83	6.13	1.51	5.87	0.10	2.16	10.33
	SD	34.38	10.61	0.55	5.61	0.03	0.59	2.25
	*H*0	Ownership**	PE	PB**	CO_2^{**}	CO_2 (normalized)***	DivYId %	P/E forward
	*T*0	2.62	−1.26	−2.05	3.84	16.10	−0.42	0.60

enterprises, operating in an environment that lacks hard constraints. Thus, the best strategy is to avoid companies with large state owned as they tend to be the biggest polluters in GHG emissions and have a lower valuation in terms of P/B ratios.

4.2 Energy Intensity Per Sales

The same methodology as above is used for the next $Escore_{Energy}$ ranking that is the energy intensity per sales. The differences between the top scoring quartile and the last scoring quartile have again been tested to see if there is statistical confidence for us to acknowledge the differences.

Table 5 presents *t*-statistics and top/last quartile stocks in China A-share market ranked based on energy per sales as our chosen environmental indicator. We analyzed 222 China A-share companies part of the MSCI EM Index. The data are as of July 2018. The P/E and P/B ratios are normalized per sector. Div. Yid % is the sector average dividend yield. Government ownership is reported in the third column. We ranked the companies based on energy per given sales normalized per industry level as well (energy industry normalized). We then calculate the difference of means to find if there is a statistically significant difference between top versus bottom quartile companies in terms of the environmental indicator and financial variables presented in the table. We find that the difference of means is significant in terms of energy usage and ownership ratio, but not for financial valuation P/E and P/B ratios. The significance of the difference of means is measured via *t*-statistic $T0$ at a confidence interval of 99%, 95% and 90% and represented with (***), (**) or (*), respectively.

The table shows that energy efficiency seems not to be correlated with valuation, but once again it is negatively related to government ownership. The higher the government involvement, the higher the energy inefficiency is.

4.3 Water Intensity Per Sales

When ranked on water intensity normalized per industry, companies ranked on top quartile spend in average 25 times the water consumed from companies' position on the bottom quartile. The difference of the mean is significant at a 99% confidence interval. Despite this strong result, there is no statistical difference between the two groups in terms of financial performance or the ownership structure.

Table 6 presents *t*-statistics and top/last quartile stocks in China A-share market ranked based on water intensity per sales as our chosen environmental indicator. We analyzed 222 China A-share companies part of the MSCI EM Index. The data are as of July 2018. The P/E and P/B ratios are normalized per sector. Div. Yid % is the sector average dividend yield. Government ownership is reported in the third column. We ranked the companies based on water intensity per given sales normalized per industry level as well (water intensity industry normalized). We then calculate the

Table 5 *T*-statistics and top/last quartile stocks in China A-share market ranked based on energy per sales

Energy intensity per sales								
Company	Industry	Ownership	Industry normalized PE	Industry normalized PB	Energy	Energy industry normalized	Div. Yid%	P/E forward
Huatai Securities Co., Ltd	Financials	13.33	1.50	1.26	54.47	5.12	2.85	7.73
COSCO Ship Dev A (HK-C)	Industrials	91.47	0.84	0.52	924.67	4.12	1.65	12.28
Air China A (HK-C)	Industrials	73.16	1.39	1.23	742.60	3.31	1.65	12.28
China East Air A (HK-O)	Industrials	85.67	1.13	1.32	729.04	3.25	1.65	12.28
Haitong Sec A (HK-C)	Financials	36.00	1.93	1.12	33.77	3.17	2.85	7.73
Shanghai Fosun A (HK-C)	Health care	1.77	1.24	1.29	104.29	2.57	0.97	-
Shanxi Taigang A (HK-C)	Materials	93.64	0.36	0.54	1105.48	2.36	1.59	11.80
China Southern Airlines Co., Ltd	Consumer discretionary	82.25	1.26	1.10	759.20	2.22	1.97	12.56
Aluminium Corp of China H	Materials	77.50	23.88	1.62	1024.13	2.18	1.59	11.80
Dong E. E. Jiao A (HK-C)	Information technology	7.76	0.57	1.33	16.80	1.81	0.77	-
Hainan Airlines Holding Co., Ltd	Consumer discretionary	0.00	1.11	0.47	558.28	1.64	1.97	12.56

(continued)

Table 5 (continued)

Energy intensity per sales								
Company	Industry	Ownership	Industry normalized PE	Industry normalized PB	Energy	Energy industry normalized	Div. Yid%	P/E forward
PetroChina Co A (HK-C)	Energy	99.09	4.55	1.02	326.81	1.61	3.33	11.20
Datang International Power Generation Co., Ltd	Utilities	74.79	1.49	0.89	7360.77	1.59	2.14	-
	Mean	56.65	3.17	1.06	1056.95	2.69	1.92	11.22
	SD	37.03	6.06	0.34	1857.98	1.03	0.70	1.79
AECC Aviation A (HK-C)	Industrials	43.21	3.47	1.39	30.91	0.14	1.65	12.28
GF Securities A (HK-C)	Financials	24.11	1.66	1.34	1.25	0.12	2.85	7.73
Weichai Power A (HK-C)	Industrials	12.18	0.59	1.12	26.02	0.12	1.65	12.28
Huadong Medicine A (HK–C)	Health care	1.63	1.04	1.83	4.55	0.11	0.97	-
BYD Co A (HK-C)	Consumer discretionary	0.00	2.77	1.59	36.39	0.11	1.97	12.56
Zhongjin Gold A (HK-C)	Materials	93.14	7.30	1.30	37.81	0.08	1.59	11.80
Orient Sec Co A (HK-C)	Financials	56.14	0.00	0.00	0.81	0.08	2.85	7.73

(continued)

Table 5 (continued)

Energy intensity per sales

Company	Industry	Ownership	Industry normalized PE	Industry normalized PB	Energy	Energy industry normalized	Div. Yid%	P/E forward
Jiangxi Copper A (HK-C)	Materials	1.99	2.48	0.75	23.05	0.005	1.59	11.80
SAIC Motor Corp A (HK-C)	Consumer discretionary	84.89	0.69	0.76	11.51	0.03	1.97	12.56
China Pacific A (HK-C)	Financials	34.92	2.88	2.44	0.19	0.02	2.85	7.73
Ping An Ins A (HK-C)	Financials	27.45	1.58	2.41	0.16	0.01	2.85	7.73
China Everbright A (HK-C)	Financials	44.73	0.71	0.69	0.05	0.00	2.85	7.73
China Life Ins A (HK-C)	Financials	3.67	3.04	2.39	0.05	0.00	2.85	7.73
	Mean	32.93	2.17	1.39	13.29	0.07	2.19	9.97
	SD	29.78	1.82	0.72	14.62	0.05	0.65	2.25
	*H*0	Ownership*	PE	PB	Energy**	Energy	Div. Yid %	P/E forward
	*T*0	1.80	0.57	−1.50	2.03	9.18	−1.02	1.57

Table 6 *T*-statistics and top/last quartile stocks in China A-share market ranked based on water intensity per sales

Water intensity per sales								
	Industry	Ownership	Industry normalized PE	Industry normalized PB	Water intensity per sales	Water intensity industry normalized	Div. Yid %	P/E forward
Top quartile companies								
China Const BK A (HK-C)	Financials	68.4	0.9	1.0	25.4	4.9	2.9	7.7
Aluminium Corp of China	Materials	77.5	23.9	1.6	36135.9	4.2	1.6	11.8
AECC Aviation A (HK-C)	Industrials	43.2	3.5	1.4	416.7	3.0	1.7	12.3
BYD Co A (HK-C)	Consumer discretionary	0.0	2.8	1.6	266.2	2.4	2.0	12.6
Shanghai Pharma A (HK-C)	Health care	31.2	0.7	0.6	763.1	2.0	1.0	-
Metallurgical A (HK-C)	Industrials	94.0	1.1	0.9	234.1	1.7	1.7	12.3
Daqin Railway A (HK-C)	Industrials	86.1	0.6	0.8	229.3	1.7	1.7	12.3
China CITIC BK A (HK-C)	Financials	92.5	0.8	0.7	8.2	1.6	2.9	7.7
ShanghaI Fosun A (HK-C)	Health care	1.8	1.2	1.3	605.8	1.5	1.0	-
	Mean	55.0	3.9	1.1	4298.3	2.5	1.8	11.0
	SD	35.2	7.1	0.4	11258.7	1.2	0.6	2.0

(continued)

Table 6 (continued)

Water intensity per sales

	Industry	Ownership	Industry normalized PE	Industry normalized PB	Water intensity per sales	Water intensity industry normalized	Div. Yid %	P/E forward
Bottom quartile companies								
Maanshan Iron A (HK-C)	Materials	93.5	0.5	0.7	1118.0	0.1	1.6	11.8
Hainan Airlines Holding Co., Ltd	Consumer discretionary	0.0	1.1	0.5	13.5	0.1	2.0	12.6
COSCO Ship Dev A (HK-C)	Industrials	91.5	0.8	0.5	15.1	0.1	1.7	12.3
Huadong Medicine A(HK-C)	Health care	1.6	1.0	1.8	41.3	0.1	1.0	-
China Pacific A (HK-C)	Financials	34.9	2.9	2.4	0.4	0.1	2.9	7.7
China Everbright A (HK-C)	Financials	44.7	0.7	0.7	0.3	0.1	2.9	7.7
Zijin Mining A (HK-C)	Materials	12.1	1.7	1.5	434.1	0.1	1.6	11.8
Weichai Power A (HK-C)	Industrials	12.2	0.6	1.1	6.5	0.0	1.7	12.3
BBMG Corp A (HK-C)	Materials	16.4	1.2	0.7	214.0	0.0	1.6	11.8

(continued)

Table 6 (continued)

Water intensity per sales

	Industry	Ownership	Industry normalized PE	Industry normalized PB	Water intensity per sales	Water intensity industry normalized	Div. Yid %	P/E forward
	Mean	34.1	1.2	1.1	204.8	0.1	1.9	11.0
	SD	34.0	0.7	0.6	351.0	0.0	0.6	1.9
	H0	Ownership	PE	PB	Water	Water (normalized)***	Div. Yid %	P/E forward
	T0	1.3	1.2	0.0	1.1	6.4	−0.2	0.0

difference of means to find if there is a statistically significant difference between top versus bottom quartile companies in terms of the environmental indicator and financial variables presented in the table. We find that the difference of means is significant in terms of water usage, but none of the financial valuation P/E and P/B ratios including the ownership ratio. The significance of the difference of means is measured via t-statistic T0 at the confidence interval of 99%, 95% and 90% and represented with (***), (**) and (*), respectively.

4.4 Waste Generated Per Sales

No significant result is produced when we rank on waste generated per sales. We believe that lack of results here is a clear outcome related to the tiny number of companies under analysis.

Table 7 presents *t*-statistics and top/last quartile stocks in China A-share market ranked based on waste generated per sales. We analyzed 221 China A-share companies part of the MSCI EM Index. The data are as of July 2018. The P/E and P/B ratios are normalized per sector. Div. Yid % is the sector average dividend yield. Government ownership is reported in the third column. We ranked the companies based on waste generated per given sales normalized per industry level as well (waste generated industry normalized). We then calculate the difference of means to find if there is a statistically significant difference between top versus bottom quartile companies in terms of the environmental indicator and financial variables presented in the table. We find that the difference of means is significant in terms of water usage, but none of the financial valuation P/E and P/B ratios including the ownership ratio. The significance of the difference of means is measured via *t*-statistic $T0$ at the confidence interval of 99%, 95% and 90% and represented with (***), (**) and (*), respectively.

4.5 Percentage of Water Recycled

Despite only a handful of companies on each quartile, there seems to be a significant relationship between recycling and P/B valuation only. There is no association with ownership, dividend yield or P/E ratios. However, the sample is too small to make further comments on this variable (Table 8).

Overall, CO_2 pollution and energy efficiency enjoy the most robust results. The least performing companies on energy and CO_2 emission (top quartile) are also the ones characterized by high government ownership and from low financial valuation.

We see no clear relationship between P/E ratios and percentage dividend payout with any of the five E indicators we selected. Our interpretation is that, as an emerging economy, China share market shows anomalies. Strict constraints, common

Table 7 *T*-statistics and top/last quartile stocks in China A-share market ranked based on waste generated per sales

Waste intensity per sales

		Ownership	Industry normalized PE	Industry normalized PB	Waste generated per sales	Waste generated industry normalized	Div. Yid %	P/E forward
China Molybdenum A (HK-C)	Materials	37.17	2.90	2.00	4511.36	3.70	1.59	11.80
Shanghai Fosun A (HK-C)	Health care	1.77	1.24	1.296	5.58	2.50	0.97	-
BOE Tech Group A (HK-C)	Information technology	31.23	0.79	0.79	1.40	1.99	0.77	-
China Intl Mar A (HK-C)	Industrials	16.37	1.71	1.33	2.98	1.70	1.65	12.28
CSSC Offshore and Marine Engineering Group	Industrials	90.70	32.06	2.15	2.48	1.41	1.65	12.28
	Mean	35.45	7.74	1.51	904.76	2.26	1.33	12.12
	SD	30.24	12.18	0.50	1803.30	0.81	0.38	0.23
Maanshan Iron A (HK-C)	Materials	93.50	0.45	0.68	179.34	0.15	1.59	11.80
Zijin Mining A (HK-C)	Materials	12.06	1.69	1.55	178.81	0.15	1.59	11.80
Shanghai Pharma A (HK-C)	Health care	31.23	0.65	0.56	0.22	0.10	0.97	-

(continued)

Table 7 (continued)

Waste intensity per sales

		Ownership	Industry normalized PE	Industry normalized PB	Waste generated per sales	Waste generated industry normalized	Div. Yid %	P/E forward
ZTE Corp A (HK-C)	Information technology	2.77	0.99	1.61	0.01	0.01	0.77	-
Sinopec Shanghai A (HK-C)	Materials	96.24	0.66	1.24	8.78	0.01	1.59	11.80
	Mean	47.16	0.89	1.13	73.43	0.08	1.30	11.80
	SD	40.03	0.44	0.43	86.32	0.06	0.36	0.00
	*H*0	Ownership	PE	PB	Waste	Waste (normalized)***	Div. Yid %	P/E forward
	*T*0	−0.52	1.26	1.29	1.03	6.03	0.10	3.16

Table 8 *T*-statistics and top/last quartile stocks in China A-share market ranked based on percentage water recycled

Company	Industry	Ownership	Industry normalized PE	Industry normalized PB	Recycled	Recycled industry normalized	Div. Yid %	P/E forward
Heilan Home A (HK-C)	Consumer discretionary	2.77	1.46	0.41	98.40	1.00	1.97	12.56
Metallurgical A (HK-C)	Industrials	93.97	1.13	0.92	18.77	1.77	1.65	12.28
Aluminium Corp of China H	Materials	77.50	23.88	1.62	96.28	1.07	1.59	11.80
	Mean	58.08	8.82	0.98	71.15	1.28	1.74	12.21
	SD	39.68	10.65	0.49	37.05	0.35	0.17	0.31
Zijin Mining A (HK-C)	Materials	12.06	1.69	1.55	87.44	0.98	1.59	11.80
China Molybdenum a (HK-C)	Materials	37.17	2.90	2.00	77.72	0.87	1.59	11.80
CSSC Offshore and Marine Engineering Group	Industrials	90.70	32.06	2.15	2.39	0.23	1.65	12.28
	Mean	46.64	12.22	1.90	55.85	0.69	1.61	11.96
	SD	32.80	14.04	0.26	38.01	0.33	0.03	0.23
	*H*0	Ownership	PE	PB***	Recycled	Recycled (normalized)***	Div. Yid %	P/E forward
	*T*0	0.38	−0.33	−2.85	0.50	2.14	1.30	1.13

***Statistically significant at 95% confidence interval

trading suspensions and segmented share classes negatively impact pricing formation and market liquidity. Additionally, preferential differences of Chinese investors, valuation and investing process shape the unique Chinese market. The concerns on environmental issues and sustainable development are still at the very early stage comparing to the growth prospects. Furthermore, limited dissemination of companies' information does not attract the attention of investors and financial analysts. Hence, the effect of environmental performance may have not substantially integrated into the firm value and may be unable to be reflected on certain valuation multiples.

5 Concluding Remarks

With the development of ESG investing across the globe and the rising heated discussion on sustainable development led by the Chinese government, the environmental factors are closely watched both domestically and internationally. At the same time, the opening up of China A-share market has created a new dimension of complexity for those who seek to understand the behavior of these factors in the Chinese market.

This paper analyzed 222 companies that are part of MSCI China A-share index. In light of increased market awareness on climate risk, we studied the relationship between specific environmental factors and company valuations. For the environmental factors, we use the following indicators: CO_2 (total GHG CO_2 emission intensity per sales), *energy* (energy intensity per sales), *water* (water intensity per sales), *waste* (waste generated per sales) and *recycled* (percentage of water recycled). To avoid sector biases and the noise in the data, we normalize the score using the industry average. Similarly, we also normalize the two financial valuation ratios under consideration, P/E and P/B ratios.

The first-level results show that certain sectors such as consumer discretionary and industrials are the most polluting in terms of CO_2 emissions (utilities not reporting on this indicator), while real estate and utilities are the least efficient in energy usage. Utilities top the rank again along materials in terms of substantial water usage. The figures in waste and recycling are more difficult to compare due to a large number of companies not reporting on these indicators.

The second-level analysis ranked the companies on our four environmental indicators, the difference between top quartile companies (the most air, water polluting and energy inefficient) to the bottom quartile companies (less air, water polluting and more energy efficient); it is always statistically significant despite the narrow sample size of 222 companies. More importantly, this difference is not driven by market capitalization (see Table 9) as one would expect and prove for ESG score in general, but a lot correlated with % government ownership in these companies.

We see a significant difference in the % government ownership and normalized financial valuation P/B ratio between high quartile of CO_2 and low quartile of CO_2. The same holds true for the energy intensity per sales.

Intuitively, SOEs incline to operate in an environment in which there are no hard budget constraints. However, on the policy front, we expect environmental law reforms will set a price on pollution and CO_2 emissions. It will introduce more costs for polluters, which is a particular concern for SOEs that historically shows 3 times higher average value of penalties, and 4% lower average net margins, than private sector peers according to the [9] report[6]. Thus, the energy efficiency cannot be ignored by SOEs anymore. Taken together with regulatory oversight over pollution across China at this time, these indicators represent a red flag. If the companies stay unchanged and operate with low efficiency, they are not likely to afford the increasing costs and would be most impacted by the environmental reforms, which should arouse investors' attention.

Our third-level analysis ranks companies according to environmental performance. It reveals significant differences between the top tier companies and bottom tier companies. In particular, when controlling per total GHG CO_2 intensity per sales and percentage water recycled, the difference in valuation as captured by P/B becomes significant. These results speak to the significance of environmental factors in company valuation in China A-share market and have implications for asset managers who have committed to the integration of environmental factors in their investment decisions. Interpretation has been offered and focused on the current market landscape of China and the investors' appetite.

On a final note, for China A-share market, the ESG and climate risk literature are scarce, and the knowledge between financial performance and ESG factors remains fragmented, on which further analysis could be conducted. Picking up on our analysis and upon the availability of more data, it would be interesting to find the impact of each of these indicators on the firm cost of equity, cost of debt and risk and finally explain any related mispricing or just the lack of pricing.

[6]https://www.msci.com/documents/10199/8b447f98-50bc-4d3d-b3f3-d4000a7084e7.

Besides data, the role of environmental indicators on company's financials strongly relates to country effect and industry characteristics. While we try to control for the later one, by normalizing it, the country effect is relevant given significantly different regulatory and market environments. For example, our expectations are that high CO_2-emitting and water-intensive companies in China should be associated with a higher cost of capital and lower valuations than otherwise. Firmly believing that as these risk factors are not priced in yet, from the asset managers and owner's perspective there is ample room for market opportunities and portfolio enhancements. This aligns well and should be the base for a smarter and more efficient funding mechanism toward a low-carbon economy.

Acknowledgements We acknowledge our research assistant RUAN, Kaiyue of HKUST 2018, for handling the data and doing the analysis on this research project.

Appendix 1

See Table 9.

Table 9 Cross-industry comparison, China A-share market, pre-normalized valuation multiples

Market cap			PE	PB	CO_2	energy	Water	Waste	Recycled
Top tier[a]	Average	374782.41	19.49	2.87	24.02	290.04	78.60	2.40	61.71
	SD	409536.78	13.81	3.02	52.70	874.31	116.82	2.34	-
Bottom tier[a]	Average	22620.78	71.75	2.67	132.19	1585.16	1235.24	60.97	2.39
	SD	5400.02	111.27	2.32	179.35	2870.07	1645.90	102.51	-
	Ho	Market cap[b]	PE[b]	PB	CO_2	energy	Water	Waste	Recycled
	T-score	6.08	−3.30	0.36	−0.85	−1.09	−1.40	−0.99	-

[a]In terms of weightings with respect to MSCI EM Index
[b]Statistically significant at 95% confidence interval

References

1. Friede G, Timo B, Bassen A (2015) ESG and financial performance: aggregated evidence from more than 2000 empirical studies. J Sustain Fin Inv 5(4):210–233. https://doi.org/10.1080/20430795.2015.1118917
2. Goss A, Roberts G (2011) The impact of corporate social responsibility on the cost of bank loans. J Bank Fin 35(7):1794–1810
3. Gupta K (2018) Environmental sustainability and implied a cost of equity: international evidence. J Bus Ethics 147(2):343–365

4. Trinks A, Scholtens B (2017) The opportunity cost of negative screening in socially responsible investing, J Bus Ethics 140(2):193–208
5. Oikonomou I, Brooks Ch, Pavelin S (2012) The impact of corporate social performance on financial risk and utility: a longitudinal analysis, financial management, 18 April 2012
6. Gonenc H, Scholtens B (2017) Environmental and financial performance of fossil fuel firms: a closer inspection of their interaction. Ecol Econ 132:307–328
7. Trinks A, Ibikunle G, Machiel M, Bert S (2018) Carbon intensity and the cost of equity capital, SSRN. https://papers.ssrn.com/sol3/papers.cfm?abstract_id=3035864. Accessed Jan 2019
8. Sanches I, Ferrero J (2017) Sustainability assurance and cost of capital: does assurance impact on credibility of corporate social responsibility information? Bus Ethics: A Eur Rev 26(3):223–239
9. MSCI (2015) China's economic transformation: a new era of ESG opportunity. https://www.msci.com/documents/10199/8b447f98-50bc-4d3d-b3f3-d4000a7084e7

Does the "Green" Attribute Affect Green Bond Pricing? An Empirical Analysis

Jingyan Fu, Jialin Lin and Jueqian Gao

Abstract In recent years, green bond has been developed progressively as a financial instrument in various capital markets. China has become the world's second largest issuer of green bonds. Based on the characteristics of China's green bond issuance, this paper examines the influencing factors of green bond issuance pricing in China. This paper constructs a multiple regression model with a credit spread as a measure of bond pricing and selects the bonds issued in the Chinese market from 2016–2018 as a sample, focusing on the mechanism and impact of the "green" attribute on bond issuance pricing. The empirical results show that: (i) the "green" nature of the bond does not affect the issue pricing but that bonds certified by a third-party green assessment are more acceptable by the market; and (ii) when the market is stable or in a rising period, green bonds embed a discount on the pricing while the investors would pay more attention to their credit risk during a market downturn. The research in this paper aims to enhance awareness of this emerging green bond market of China and urges the government to improve its guidance for investing in green debt market, which has meaningful policy implications.

Keywords Green bond · Credit spread · Issue pricing · "Green" attributes

1 Introduction

China is currently in a critical period of ecological, civil construction, and there is strong demand for funding for economic green development. At the end of 2015, the People's Bank of China issued the "Announcement on Issues Related to the Issuance of Green Financial Bonds in the Inter-bank Bond Market" (People's Bank of China Announcement [4] No. 39), launched green financial bonds and officially launched the construction of China's green bond market. Subsequently, the National

J. Fu (✉) · J. Gao
Institute of Resource, Environment and Sustainable Development, College of Economics, Jinan University, Guangzhou, China
e-mail: tfjyan@jnu.edu.cn

J. Lin
Centre Branch of the People's Bank of China, Zhuhai, China

J. Fu and A. W. Ng (eds.), *Sustainable Energy and Green Finance for a Low-carbon Economy*, https://doi.org/10.1007/978-3-030-35411-4_10

Development and Reform Commission, the China Securities Regulatory Commission, the Interbank Market Dealers Association, the Shanghai Stock Exchange and the Shenzhen Stock Exchange successively issued a series of documents to promote the comprehensive development of China's green bond market in order to meet the funding needs of green projects and support the development of the green economy. It plays a pivotal role.

Compared with ordinary bonds, green bond fundraising projects can yield good environmental benefits, but these need to improve their information disclosures on the use of the funds raised and the related environmental benefits. Additionally, the certification assessment is paid by the issuer, which enhances the comprehensive distribution of green bonds to a certain extent. Furthermore, the environmental benefits consequently generated by green bonds, the related information disclosures and evaluation and certification results can provide investors with investment decision-making references, which in turn enable investors to control environmental risks linked with the investment returns. Therefore, whether investors are willing to adjust the risk premium requirement for the "green" attribute of bonds and achieve a "win-win" regarding the environmental and economic benefits (in terms of pricing) is crucial to promoting the sustainable and healthy development of the green bond market.

By examining the green bonds issued in China in 2016–2018, this paper empirically tests the factors influencing green bond issuance pricing through multiple regression models. The study analyses the reasons behind these factors and proposes relevant policy recommendations. The study results help to refine the relevant supporting policies based on the characteristics of green bonds and to establish a market-oriented, long-term development mechanism. This study has certain reference value for improving the green bond issuance pricing mechanism and for enhancing the attractiveness of green bonds and the vitality of the green bond market.

2 Status of the Development of China's Green Bond Market

The European and American green bond markets began in 2007–2008, with the issuance of green bonds by the World Bank and the European Investment Bank. The Chinese green bond market began to develop in 2016 with the issuance of the first green financial bonds by the Shanghai Pudong Development Bank in the interbank market. Although the Chinese green bond market started late, China has played an important role in promoting the development of the global green bond market. In 2016, the first year of China's green bond market, China, became the world's largest issuer of green bonds, amounting to 19.83 billion yuan, or 44.69% of the global green debt market. In 2017, China's green bond issuance exceeded 200 billion yuan for the first time, reaching 206.78 billion yuan, which accounted for 15.11% of the world's green debt issuance, second only to the USA. In 2018, China's green bond issuance amounted to 220.53 billion yuan (a year-over-year increase of 6.56%), which accounted for 18.65% of the global green debt issuance (a year-over-year increase of

3.54%), still ranking second in the world. In the past three years, China's green bond market has developed steadily, and its activity has continued to increase. China's green bond market has been characterized by a diverse variety of bonds, good credit quality of issuers, a high proportion of bonds with green evaluation and certification and good environmental benefits from fundraising projects.

2.1 Growth of Green Bonds in the Capital Market of China

As the first product of China's green bond market, green financial bonds play an important role in promoting market development. With the introduction of a series of policies by relevant regulatory authorities, the China Interbank Bond Market, Shanghai Stock Exchange and Shenzhen Stock Exchange have been successively launched. Relevant green bond products, including green financial bonds, green corporate bonds, green debt financing instruments, green panda bonds and green asset securitizations, can effectively meet the diverse financing needs of enterprises. The participation of financial institutions in the green bond market has increased. In terms of the amount of issuance, 128.92 billion yuan in green financial bonds was issued in 2018, a decrease of 26.08 billion yuan from 2016, and China's share of the global green bond market fell from 78.07% in 2016 to 58.51% in 2018. Meanwhile, financial bonds amounting to 91.433 billion yuan were issued in 2018, which is an increase of 110.05% over 2016. There were 128 green bonds were issued in 2018, 90 of which were non-financial bonds, which is an increase of 13.17% over 2016.

2.2 Corporate Green Bond Issuers on a Large Scale

With the promotion and application of green bonds, increasing numbers of financial institutions and enterprises have issued green bond financing, and the participating entities have gradually diversified. In 2016–2018, a total of 174 entities issued green bonds, including 38 green bond issuers in 2016. In 2017, there were 73 green bond issuers and 67 new issuers. In 2018, there were 99 green bond issuers and 69 new issuers. From the perspective of ownership, there were 123 state-owned enterprises among the green bond issuers in 2016–2018, accounting for 70.69%. There were 23 listed companies, accounting for 13.22%, and 22 private enterprises, accounting for 12.64%. From an industry perspective, a total of 53 financial institutions issued green bonds in 2016–2018, accounting for 30.46% of the issuers; 121 non-financial companies issued green bonds, accounting for 69.54% of the issuers. In terms of geographical distribution, there were 99 (issuers from economically developed regions such as the eastern coast, accounting for 56.9%, and 75 issuers from other regions, accounting for 43.1%.

2.3 Green Bond Ratings

China's green bond market is still in the initial stages of development with a limited number of issuers and aggregates, and the fundraising projects are mostly supported by government policies. Therefore, the green bond ratings are better than the overall level of the Chinese bond market. At the same time, as acceptance of the green bond market increased, low-rated issuers and debt were gradually accepted by investors. Of the green bonds issued in 2018, 39.08% had *AAA* ratings, which was 19.22% lower than in 2016, and 4.34% higher than the overall bond market. The *AAA* debt rating accounted for 51.82%, which was a decrease of 5.28% from 2016 and 2.25% higher than the overall bond market. In 2018, the minimum subject rating and debt rating both sank to A^{+}. In 2016–2018, a total of 39 green bonds issued by 30 companies increased their debt rating by adding third-party guarantees or by providing collateral.

2.4 Certified Green Bonds on the Rise

With the development of green bonds, the number of green bonds that have obtained third-party green certification has increased, but the proportion has declined. At the same time, the participation of various types of institutions has promoted the diversification of China's green assessment and certification methods, and the evaluation system has gradually improved. In 2016–2018, a total of 207 green bonds received third-party green certification, accounting for 67.8%. Of this total, 80 green bonds were awarded third-party green certification in 2018, 38 more than in 2016 and 4 more than in 2017. This number accounted for 62.5% of all green bonds issued that year, which was down 23.2% from 2016 and down 3.59% from 2017. Since 2016, 15 different types of third-party organizations have participated in green bond evaluation and certification work, including accounting firms, professional rating and certification institutions, energy and environment consulting organizations, and other academic institutions. The top three organizations conducting green evaluation and certifications are the Yongan Accounting Firm, the China Bond Credit Rating Co., Ltd., and the China Energy Conservation Consulting Co., Ltd. The Yongan Accounting Firm mostly assesses the green bond principles prior to issuance of green bonds or the certification of green loan principles. China Bond Ratings issues bond ratings and conducts follow-up assessments of green projects. China Energy Conservation mainly conducts project analyses and provides specific opinions. The green assessment and certification method is scientific, and the certification process is continuously improved. Additionally, green bonds are continuously adapting to domestic and international issuance standards.

2.5 *Green Bonds Supporting a Wide Range of Areas and Clean Energy Projects*

In 2016–2018, the green bonds issued by different types of businesses in China have covered the six categories of green projects in the Green Bond Support Project Catalogue (2015 Edition), indicating a broad diversity in the types of projects. In 2016–2018, non-financial green projects raised a total of 18.038 billion yuan, of which clean energy projects accounted for the highest proportion (approximately 36.38%). Of the remaining funds, 16.64% were invested in clean transportation projects, 14.9% in pollution prevention and control projects, 14.05% in energy conservation projects, 10.12% in resource conservation and recycling projects and 5.1% in ecological protection and climate change adaptation projects. In addition, 2.83% of the funds raised could not be categorized due to unclear information in the disclosures.

3 Research Design: Factors Affecting the Pricing of Green Bond Issuance in China

Multiple regression models are commonly used to study bond pricing issues. Merton [3] proposed the credit risk structure model of securities pricing based on the option pricing model proposed by Black and Scholes [1]. Jarrow and Turnbull [2] priced bonds with a risk-free part and a default risk part and established a simple model of credit risk, which further enriched securities pricing theory. Relevant research shows that bond pricing is ultimately reflected in the risk–reward ratio required by investors, and the multiple-regression model can be used to analyse bond pricing.

When analysing the green bonds issued in 2016, Yao Minglong proposed that green bonds should have a discount on the issue rate. Empirical analysis proves that issue size, issue period, registered capital, return on net assets and market interest rates affect the discount rate for green bond issuance interest rates. Wang Xiaoling used the 69 green bonds issued in 2016 and 2017 as a sample. The regression model proved that the benchmark interest rate, rating and third-party certification significantly affected the pricing of green bond issuance. The bond maturity had no significant impact on the pricing of green bond issuance. Research by domestic scholars on green bond pricing mainly uses the issued green bonds as a sample and analyses its influencing factors through multiple regression models. There is little comparison between green bonds and ordinary bonds, and little analysis of the impact of the "green" attribute of green bonds on the pricing of offerings.

Based on the existing domestic and international research, this paper uses the risk–return ratio (credit spread) to measure the pricing level of bonds and further breaks down the influencing factors of credit spreads into default risk (credit risk), expected liquidity risk and bond features, and selects the corresponding indicators to construct a multiple regression model. The bonds issued in the Chinese market in

2016–2018 were selected as samples, focusing on the mechanism and impact of the "green" attribute on bond issuance pricing.

3.1 Definition of Spreads

Considering that China's bonds are mainly issued at low prices and that domestic and international researchers all use government bond yields as a risk-free bond yield, the formula for bond issuance pricing is simplified as: the issue price (the coupon rate R) = the national debt yield (R_{ND}) + credit spread (ΔR). Considering that the credit spread can reflect the value of a bond investment, this influencing factor is also the core factor in determining the bond issuance price. Therefore, this paper uses the credit spread as a dependent variable to measure the bond issue price. According to the principle of matching time and term, we compare the bond yield curve (expiration) of the China Bond and obtain the credit spread of each bond (ΔR).

$$\Delta R = R - R_{\text{ND}}$$

Among these variables, R_{ND} represents the same or similar period of government bond yield on the day of each bond issue. That is, if there is no bond yield for the same period, this value is calculated according to the government bond yield with the adjacent longer term. For example, the government bond yield corresponding to the 2-year bond is based on the 3-year bond yield on the day of issue.

3.2 Model Design

Based on theoretical analysis and practical experience, we construct a multiple regression model to estimate the credit spread of bonds:

$$\Delta R = \beta_0 + B_1\text{CR} + \beta_2\text{IS} + \beta_3\text{DM} + \beta_4\text{GB} + \beta_5\text{TPC} + \varepsilon$$

The independent variables in the model are as follows:

i. Subject rating (CR), qualitative indicators, drawing on the common practice at home and abroad, assigning the bond issuer's subject rating to $AAA = 1$, $AA^{+} = 2$, $AA = 3$, $AA^{-} = 4$, $A^{+} = 5$. A larger value indicates a lower rating.
ii. The issuance scale (IS) and quantitative indicators are the actual issuance amount of the bonds.
iii. The bond term (DM), the quantitative indicator, is the actual term of the bond and is standardized in "years".
iv. Whether it is a green bond (GB): a qualitative indicator, "yes" is taken as 1 and "no" is taken as 0.

v. Whether a third-party green certification (TPC) is obtained: a qualitative indicator "yes" is taken as 1 and "no" is taken as 0.

3.3 Research Hypothesis

The purpose of this paper is to explore the impact of the "green" attribute of green bonds on the bond issuance price and to combine the practical experience and related literature results to propose the following construction for empirical testing.

Hypothesis 1: Credit spread is positively correlated with bond issuer credit risk. This paper uses subject rating (CR) as an important indicator to measure the issuer's credit status. The poorer the ability of the issuer to perform the subject rating is, the higher the risk return required by the investor is.

Hypothesis 2: Credit spreads are positively correlated with expected liquidity risk. This paper uses the issue size (IS) and bond maturity (DM) as an important indicator to measure the expected liquidity risk. The larger the scale of issuance is, the wider the range of investors holding the bond is. The higher the liquidity of the bond and the stronger the issuer's ability to finance are, the more investors are willing to accept lower credit spreads. From the perspective of bond maturity, due to the long investment cycle of green projects, the longer the bond term is, the higher the liquidity matching of green bond investment projects is. The lower the liquidity risk is, the more investors are willing to accept lower credit spreads.

Hypothesis 3: Credit spreads are inversely related to the "green" attributes. This paper selects third-party green certification (TPC) of green bonds (GB) as an important indicator for measuring the "green" attributes. According to the previous analysis, China's green bond market is dominated by financial institutions and state-owned enterprises. Green bond ratings are better than the overall market level, and green bonds are strongly supported by state and local governments. Therefore, investors are willing to reduce credit spreads for green bonds. The third-party green certification endorses the "green" nature of the bond, and its issuer's qualifications are generally better than that of an uncertified green bond issuer. Bonds with green certifications have clearer environmental risks and benefits of the investment project. Information disclosure is better and can alleviate information asymmetry. Therefore, investors are willing to accept lower credit spreads for bonds with third-party green certifications.

3.4 Sample Selection

This paper selects corporate bonds, debt financing instruments and financial bonds issued in the Chinese bond market in 2016–2018 as research samples. The data come from the Wind database and are screened according to the following criteria. First, for sample comparability, the green and non-green bonds (of the same type) issued on the same day during 2016–2018 were selected as samples, and the remaining bonds were deleted. Second, taking into account the differences in bond pricing systems, policy-based financial bonds and asset securitization products were eliminated. Third, taking into account the integrity of information disclosure, private debt and targeted financing tools were eliminated. Fourth, since only five green, short-term financing bonds and green panda bonds were issued in 2016–2018, these were excluded in order to avoid the impact of extreme data. Fifth, bonds without subject ratings were eliminated.

To observe the impact of the "green" attribute on the bond issue price during the bond market's rising, stationary and down periods (per the China Bond China Green Bond Index), the overall sample was processed and analysed. The bonds issued from November 2016 to January 2018 are classified as market down-period bonds; the bonds issued from January–October 2016 and February–December 2018 are classified as market upswings or stationary bonds.

4 Analysis of Empirical Results

Using the Stata software, we find that there is a weak correlation between the explanatory variables affecting ΔR, so it is suitable to use them to analyse the contribution of each of the explanatory variables ΔR. The empirical tests led to the conclusions presented below.

4.1 The "Green" Attribute of Bonds Does not Affect the Issue Pricing, and Bonds with Third-Party Green Certifications Are Better Accepted by the Market

From the multivariate regression results in Table 1, it can be seen that the main body rating, total issuance and bond maturity passed the 1% significance-level test,

Table 1 Overall results from the sample regression analysis

Risk premium	Coefficient	Standard error	T value	P value
Subject rating	0.190897	0.009166	20.83	0.000
Total issue	−0.00712	0.000649	−10.96	0.000
Bond term	−0.05943	0.006961	−8.54	0.000
Whether green bonds	−0.06276	0.099478	−0.63	0.528
Whether third-party green assessment certification	−0.4391	0.137165	−3.20	0.001
Observations	5485			

indicating that the credit risk of the bond and the expected liquidity risk have a significant impact on the credit spread. Among these factors, the subject rating is positively related to the credit spread, indicating that the higher the subject rating is, the lower the risk premium required by investors is. The issue size, bond maturity and credit spread are negatively correlated, indicating that the greater the expected liquidity risk is, the higher the risk premium demanded by investors is. The empirical results are basically consistent with Hypothesis I and Hypothesis II.

However, the coefficient of green bonds did not reach a 1% level of significance. There is no strong correlation between green bonds and credit spreads, indicating that investors will not reduce risk premium requirements for green bonds. The empirical results are inconsistent with the assumptions for three green bonds. There is a negative correlation between third-party green certification and credit spreads, indicating that market investors are more willing to accept green bonds with third-party certifications to appropriately reduce risk premium requirements.

4.2 In a Market that Is Stable or Rising, Green Bonds Have a Discount on the Issue Rate. in a Down Period for Markets, Investors Pay More Attention to Credit Risk

The regression results for stable, rising and declining bond markets are presented in Table 2. When the bond market is stable or rising, the main body rating and the total issuance pass the 1% significance level test. The subject rating is positively related to credit spreads, and the total amount of issuance is negatively correlated with credit spreads, which is basically consistent with the regression of the overall sample. The bond maturity did not pass the significance test, and investor attention to the bond maturity declined during this period.

Unlike the overall sample regression results, the bond market is stable or rising. Whether a bond is "green" passed the significance level test (at the 1% level) and is negatively correlated with the credit spread when it has a third-party green certification. The impact of the spread is not significant, indicating that investors have higher

Table 2 Sample regression analysis results for Jan–Nov 2016 and Feb–Dec 2018

Risk premium	Coefficient	Standard error	*T* value	*P* value
Subject rating	0.172886	0.012958	13.34	0.000
Total issue	−0.00516	0.000824	−6.27	0.000
Bond term	−0.00149	0.010037	−0.15	0.882
Being qualified as "green" bond	−0.38721	0.129561	−2.99	0.003
Whether third-party green assessment certification is obtained	−0.1665	0.171972	−0.97	0.333
Observations	1891			

Table 3 Regression analysis results of samples from Dec 2016 to Jan 2018

Risk premium	Coefficient	Standard error	T value	P value
Subject rating	0.193341	0.01211	15.96	0.000
Total issue	−0.0085	0.000911	−9.33	0.000
Bond term	−0.08453	0.0091	−9.29	0.000
Whether green bonds	0.087343	0.137335	0.64	0.525
Whether third-party green assessment certification	−0.56257	0.194907	−2.89	0.004
Observations	3594			

acceptance of green bonds during this period, and green bonds have a discount on the issue rate. Investor demand for "green" information disclosure on green bonds has declined, weakening the investment behaviour of "excellent selection" in green bonds.

From the regression results in Table 3, it can be seen that during a period of market decline, the regression results are basically consistent with the overall sample regression results, and the subject rating and the third-party green certification have higher coefficients, indicating that credit spreads are more sensitive to the above factors during a market decline. Investors pay more attention to bond credit risk and green information disclosure during a market decline and choose green bonds with better investment qualifications.

5 Concluding Remarks

5.1 Current Status

The research results in this paper show that, first, at present, China's green bond market is still in an initial stage of development. When investing in green bonds, investors still follow the traditional approaches for investing in ordinary bonds. The investment process focuses on bond credit risk and liquidity risk. Green bonds would derive the overall bond risk by potentially reducing environmental risks. However, the advantage of risk reduction has not been effectively recognized by investors.

Second, there is a "preferential" investment phenomenon in the process of investing in green bonds. Third-party green certification can enhance the credibility of green bonds, improve the transparency of green information disclosure and provide important investment decision-making references for investors lacking environmental expertise.

Third, the market acceptance of green bonds is affected by bond market fluctuations. When the market is rising or stationary, investors are more willing to invest in green bonds and to moderately lower the risk premium requirement. The green bond "selected" investment phenomenon is weakened. Investors are concerned about risks, expected liquidity risk, etc. and are reluctant to reduce risk premium requirements for green bonds. Investors have strengthened their green bond "preferred" investment behaviour.

5.2 Recommendations

5.2.1 Cultivating Professional Green Investors and Increasing Market Awareness of Green Bonds

The first recommendation is to improve publicity and training regarding green bonds, raise the stakeholders' awareness of green bonds and enhance investor awareness of green investment. The second recommendation is to support the establishment of green bond asset management products. The funds raised must be used to invest in green bonds and provide some support for the establishment of an approval process for green asset management products to increase the enthusiasm of financial institutions for participation in green bond investment. The third recommendation is to give certain tax benefits to income from green bond management products and investment in green bonds and to reduce the cost of investing in green bonds. Fourth, the green investment-related indicators should be used as reference indicators for the supervision and evaluation of financial institutions, such as fund companies, banks and insurance companies, and encourage financial institutions to increase investment in green bonds. The fifth recommendation is to accelerate the opening of the bond market and to introduce foreign, professional green bond investment institutions to stimulate demand in the green bond market.

5.2.2 Improving the Green Bond Certification System and Its Credibility

First, provide government subsidies for green certification fees to encourage more green bonds to obtain the certification, improve the market acceptance of green bonds and thereby reduce the cost of green bond financing. The second recommendation is to study and establish green certification standards to enhance the credibility of the certification, improve the efficiency of the certification workflow and reduce the certification cost. The third recommendation is to strengthen supervision, standardize the development of green assessment and certification institutions, enhance self-disciplined management, promote the green evaluation and certification market and improve the independence, effectiveness, scientific and objectivity of green assessment certification.

5.2.3 Establishing a Green Bond Credit Mechanism to Mitigate the Impact of Market Volatility on Green Bond Financing

First, local governments (with conditions) can pilot the establishment of a green bond financing guarantee fund, which will provide a certain amount of risk-sharing for the green bonds issued by enterprises within the jurisdiction, and release the credit risk of green bonds. Second, local governments can encourage state-owned financing guarantors to provide guarantee services for green bonds and increase the green bond rating through credit enhancement measures, thereby reducing the cost of green bond financing. The third recommendation is to develop a green bond credit risk derivatives market, encourage financial institutions to create credit risk mitigation and risk protection tools for green bonds, achieve green bond credit risk and investor risk preferences, and provide a comprehensive protection mechanism for green bond financing. These measures will help mitigate the impact of market volatility on the issuance of green bonds.

References

1. Black F, Scholes M (1973) The pricing of options and corporate liabilities. J Polit Econ 81(3):637–654
2. Jarrow RA, Turnbull SM (1995) Pricing derivatives on financial securities subject to credit risk. J Fin 50:53–85
3. Merton RC (1974) On the pricing of corporate debt: the risk structure of interest rates. J Fin 29:449–470
4. People's Bank of China (2015) Green financial bonds announcement 15–12

Financing Water Treatment in Guangdong Province Based on the PPP Mode

Jingyan Fu and Saisai Guo

Abstract The establishment of comprehensive water treatment system is an important part of China's recent urbanization and ecological civilization construction. The extreme lack of water resources, the serious pollution of water sources, as well as the high risk and complex technical requirements determine the salience and urgency of investing in water treatment infrastructures. At the same time, the water treatment projects generally require large investment amounts and extended investment recovery periods. Both construction and operations of these projects are considered complicated. Therefore, innovative water treatment investment and financing mode are imperative. In this context, the use of private capital to alleviate the financial pressures of local governments and the new financing mode of multi-participation in environmental governance is advocated. Public–private partnership (hereinafter referred to as PPP) has particularly been receiving more attention in recent years. The main purpose of this paper is to analyze the effectiveness of PPP financing methods in the implementation of water treatment in Guangdong Province and to exemplify the PPP mode as a reference for the healthy development on the basis of government guidance, social participation and market operation. It demonstrates an alternative approach to achieving multi-dimensional governance of water and promoting the sustainable development of both the economy and society in Guangdong Province. The overall infrastructure for sustainable energy development can be complemented by building up water treatment infrastructures that foster sustainability.

Keywords PPP mode · Social capital · Pluralistic governance · Environmental investment and financing

---

J. Fu (✉)
Institute of Resource, Environment and Sustainable Development, College of Economics, Jinan University, Guangzhou, China
e-mail: tfjyan@jnu.edu.cn

S. Guo
Industrial & Commercial Bank of China Guangdong Branch, Guangzhou, China

J. Fu and A. W. Ng (eds.), *Sustainable Energy and Green Finance for a Low-carbon Economy*, https://doi.org/10.1007/978-3-030-35411-4_11

1 Introduction

The rapid development of Guangdong's economy has made environmental pollution increasingly serious, mainly including air pollution, water pollution and soil pollution. Guangdong Province is located in South China, and its northern region is subtropical. The southern region has a tropical climate with abundant precipitation and abundant light and heat resources. According to statistics, there are about 600 main streams and tributaries in Guangdong, including Zhujiang, Hanjiang, Moyangjiang and Jianjiang. Although Guangdong as a province has a relatively sophisticated system of atmospheric governance, such as the promulgation of the "ten gas regulations" and the establishment of the "carbon trading" system, the water treatment facilities are still in the exploratory experimental stage and have not formed a mature management system.

Traditionally speaking, the environmental governance system is based on administrative control, and the market is auxiliary. In addition, because the parties pay different attentions in environmental governance, they will inevitably prefer their own preferences in operational management. It is difficult for the government to effectively balance the interests of all parties while taking more risks in the governance process. The introduction of PPP, a new financing method in public facilities construction projects, arranges financing according to the expected income, assets and government support measures of the project, which not only conforms to China's macroeconomic policies, but also enables more social capital to actively participate in the project. It can also absorb private enterprises, joint ventures and individuals to contribute more advanced and efficient management methods and technologies to the construction of public facilities, to achieve the long-term goal of sharing risks and sharing benefits between the government and the society, and to better serve the society.

The water treatment industry generally requires large investments, high fixed costs and long payback periods, so there are also very strict requirements on capital operation capabilities. The PPP mode is used in industries with high investment, high cost and long return on investment. The water quality improvement project in Guangdong Province can replace the traditional fund supply mode with PPP mode. The financing advantage reflects the development of multi-common governance. The path also provides new ideas, new directions and new impetus for water treatment in Guangdong Province.

2 Literature Review

Abednego and Ogunlana [1], Lee and Schaufelberger [4], Teller and Kock [11] use different empirical analysis methods to study investment risk issues in PPP projects. Mansor and Rashid [5], under the premise of affirming the Private Finance Initiative (PFI), combined with the PFI contract background, critically reviewed the integrity of the PFI contract, analyzed the causes of the incomplete contract, and proposed relevant countermeasures. Shrestha et al. [8] also pointed out through the differential analysis that there is a big gap between the government and the private sector in the sensitivity of PPP project financing risks, and the differentiation mechanism should be developed accordingly when evading project risks. Moszoro and Gasiorowski [6] based on the Pareto optimal theory, through the analysis of the cost and financing structure of PPP projects, combined with practical cases to provide a reference for the development of PPP.

In terms of case studies, Siemiatycki [9] summarized the PPP project in the UK from 1987 to 2009 and found that the implementation of the PPP project has both positive and negative effects: On the one hand, the parties involved in the PPP project have cooperated many times. Conducive to reducing transaction costs can also encourage innovation; on the other hand, the more stable trading relationship between the participants in the market helps to reduce competition, which in turn leads to increased costs and quality. Schuyler House [7] provides a powerful guide to how to strengthen the quality of regulation of rules and relationships through case-by-case analysis. Tawalare and Balu [10] selected two polar water supply projects, one under public–private partnership (PPP) and the other under direct government funding, compared to government-funded. The continuous water supply project performed well, and it is recommended to strengthen the management of PPP projects. Kim and Choi [3] explored the expected benefits of large-scale construction projects based on contractual guarantees through an analysis of the implementation experience of the New Matsushima City project in Korea (known as the world's largest private sector PPP project). The project was found to be slightly less effective than its theoretical assumptions, and project implementation was more challenging for the private sector. Fuest and Haffner [2] analyzed the impact of policy changes on the financial, technical and managerial aspects of Ghana's semi-state water company (GWCL) participation in the construction of rural water supply PPP projects.

From the above-mentioned literature on the theoretical research and practical experience analysis of the PPP mode, it can be seen that the PPP mode is involved in all walks of life in the field of public infrastructure, and some areas have been more maturely applied, but for PPP in water There is a lack of special research on technical and management requirements in environmental governance.

3 Application Status of PPP Mode in Water Treatment

3.1 Application Status of PPP Mode in China's Water Treatment

3.1.1 Large Investment Scale of PPP Projects

Since the introduction of PPP into China's urban construction, the PPP financing mode has developed rapidly in the construction and operation of China's infrastructure. According to the data provided by the PPP Integrated Information Platform, from 2012 to 2017, there are 94 PPP projects for the comprehensive management of water pollution that have been completed and under construction or operation nationwide, with a total investment of 209 billion yuan. These PPP projects are mainly distributed in relatively densely populated areas such as Beijing, Shandong, Henan, Guizhou, Anhui and Guangdong. The number of projects and investment accounted for 63.8% and 60.2%, respectively. Judging from the funds invested in the project, among the 94 PPP projects mentioned above, projects with an investment of over 100 million yuan accounted for more than 97%, mainly concentrated in the field of comprehensive river basin management, except for two pure river channel management projects with an investment of less than 100 million yuan. In addition, the average investment of other PPP projects exceeds 2 billion yuan.

3.1.2 Growing PPP Project Operations

There are three batches of PPP projects listed in the list of PPP demonstration projects after the organization of the Ministry of Finance from 2014 to 2015. There are 752 projects, of which municipal public utilities have the largest number of projects, about 00; science and technology projects have the least, less than ten. The distribution of demonstration projects in seven regions of the country is as follows: First, among all the demonstration projects in the warehouse, Huazhong, East China and Southwest China have the largest number of projects, 153, 152 and 123, respectively. Second, there are 45 in Northeast China, 186 projects involving water construction, and relatively more water projects in East China, Central China and Southwest China. There are 40 in East China, 34 in Central China and Southwest China. There are only seven water projects in Northeast China.

3.2 *Application Status of PPP Mode in Water Treatment of Guangdong Province*

According to documents issued by the Guangdong Provincial Department of Finance, in 2014, in the third wave of the development of China's PPP mode, Guangdong Province officially launched the PPP project. In practice, the PPP mode is being continuously accelerated to the infrastructure construction projects in Guangdong Province.

In the PPP mode, the investment is mainly concentrated on bridges, roads and other projects. The investment in water treatment has gradually increased since 2015. This phenomenon can be seen from the statistics of the first three batches of PPP demonstration projects in the water treatment. It can be seen as follows: First, among the first three batches of demonstration projects announced by the Ministry of Finance, the first batch does not include water treatment projects. The main reason for this phenomenon is that the PPP mode has not been officially introduced into the water treatment field in Guangdong Province at the initial stage of the demonstration project library. Second, in 2015, a total of four PPP projects in the province were selected for the second batch of PPP demonstration projects of the Ministry of Finance. These water treatment projects are mainly concentrated in the construction of water supply facilities. Maoming City and Jiangmen City of Guangdong Province, respectively, put in a water supply project. The investment in the water supply project in the Shuiwan District of Maoming City was relatively large, totaling 1.137 billion yuan. The PPP water treatment project in Guangdong Province was gradually developed. Third, in 2016, a total of 18 projects in Guangdong Province were selected into the third batch of PPP demonstration projects of the Ministry of Finance, and the investment amount has also increased greatly. Among them, four cities were involved in the field of water treatment and PPP demonstration in the field of water treatment. These projects were mainly concentrated in Shantou City (3), Jiangmen City, Maoming City and Meizhou City (one in each of the three cities). This shows that the development of PPP mode in various regions of Guangdong Province is not quite balanced (Fig. 1).

In order to understand more clearly the development of PPP projects in water treatment in Guangdong Province, Table 1 details the implementation areas, project names and project investment amounts of the demonstration projects in the water treatment in Guangdong Province mentioned above. It can also be seen that in the water treatment of Guangdong Province, the PPP projects are mainly concentrated in Shantou City. The main projects are the construction of sewage treatment plants and also involve water conservancy projects.

The increasingly mature development of PPP mode in China has made it play an increasingly important role in the field of water treatment. Therefore, the state has continuously introduced new policies to support and encourage the development of PPP projects, including the implementation of PPP projects, the establishment of supporting financial services, the improvement of supervision mechanisms, and the design and maintenance of the overall operational management system of the project.

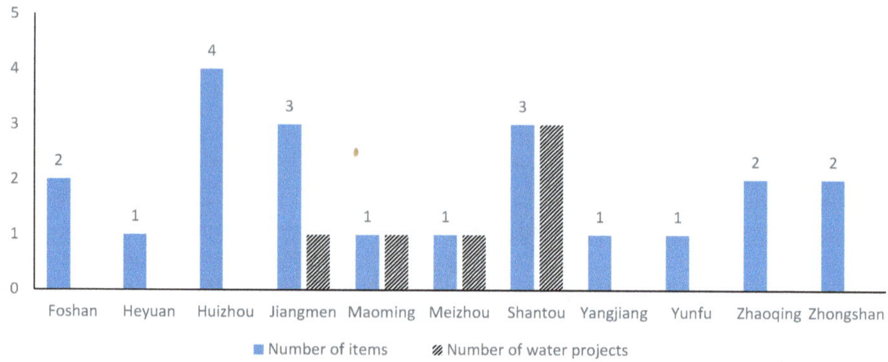

Fig. 1 Distribution of PPP demonstration projects in Guangdong Province. *Source* List of PPP demonstration projects of the Guangdong Provincial Government and the Social Capital Cooperation Center

3.3 Policy Analysis on the Introduction of PPP Mode in the Water Treatment

The increasingly mature development of PPP mode in China has made it play an increasingly important role in the field of water treatment. Therefore, the state has continuously introduced new policies to support and encourage the development of PPP projects, including the implementation of PPP projects, the establishment of supporting financial services, the improvement of supervision mechanisms and the design and maintenance of the overall operational management system of the project.

Table 2 summarizes the documents related to the use of PPP mode to participate in water treatment, mainly reflecting the following three points: First, China's application of PPP mode in the water treatment began in 2015, and it has a deep implication since the introduction. The release of several official documents has introduced the PPP mode for water treatment projects. Second, in 2016, the country's application of PPP mainly focused on soil pollution and old-age services, and there were few policies and regulations on water treatment for operating PPP mode. Third, although the state has actively introduced regulations and regulations to introduce social capital and promote environmental pluralism, there are no policies that specifically address PPP projects in the water treatment.

The development of municipal projects under the PPP mode is inseparable from the support and supervision of local government policies and regulations. Effective policies are a powerful tool for the successful operation of PPP projects. The Guangdong Provincial Government attaches great importance to the PPP project. After the establishment of the Guangdong Provincial Government and the Social Capital Center in 2014, it has also issued relevant documents to encourage the implementation, support and supervision of the PPP mode to participate in water treatment.

Table 3 summarizes the documents on the application of PPP mode in water treatment, which mainly reflects the following three information: Firstly, the application

Table 1 First three batches of demonstration projects for water treatment in Guangdong Province announced by the Ministry of Finance

Number	Area	Project	Total investment (100 million yuan)
1	Maoming City	Water supply project of Shuidongwan City, Maoming City	11.37
2	Jiangmen	Jiangmen urban emergency standby water source and water supply facilities project	2.75
3	Chaonan District, Shantou City	Chaonan, two British grids and one sewage treatment plant construction and renovation marked a mention	4.51
4	Chaonan District, Shantou City	Chaonan District three sewage treatment plants	6.80
5	Chaoyang District, Shantou City	Three sewage treatment plants in Chaoyang District, Heping, Tonglu and Guanyu	8.76
6	Meizhou City	Guangdong Hanjiang Gaochun water control project	6.16
7	Yunan County, Yunfu City	Junan County whole county domestic sewage treatment bundled PPP project	5.03

Source List of PPP demonstration projects of the Ministry of Finance and the Government and Social Capital Cooperation Center in 2015 and 2016

of PPP mode in water treatment field in China started in 2015 and has been paid more attention since its introduction. Many documents published continuously mentioned the introduction of PPP mode in water treatment projects. Secondly, in 2016, the application of PPP was mainly concentrated on soil pollution and pension services, and there were few policies and regulations on water treatment of PPP operation mode. Third, although the state has also actively promulgated regulations and regulations to improve the introduction of social capital and promote environmental pluralism, there is no specific policy or policy enactment for PPP projects in the field of water treatment.

Table 2 Document of the national level PPP mode for water management in 2015–2017

Release time	File name	Main content
April 2015	Water pollution prevention action plan	Promote diversified financing and guide social capital investment. Promote equity financing, project income rights, franchise rights and other pledge financing guarantee methods and expand financing channels
April 2015	Implementation opinions on promoting government and social capital cooperation in the field of water pollution prevention and control	Require all levels of government, financial departments and environmental protection departments to strengthen organizational management, actively coordinate and coordinate and establish an effective monitoring mechanism for PPP projects
October 2015	Opinions of the general office of the state council on promoting the construction of Sponge City	Further promote government and social capital cooperation, franchising and other modes, encourage and absorb the construction of social capital investment sponge city
October 2016	Opinions on further encouraging and guiding private capital to enter the urban water supply, gas, heating, sewage and waste treatment industry	Encourage private capital to participate in the construction and operation of municipal public facilities through government and social capital cooperation mode, and do a good job in supporting financial services for municipal project PPP mode
June 2017	Guiding opinions on innovating the institutional mechanism of rural infrastructure investment and financing	Encourage the introduction of social capital into rural infrastructure construction and encourage local governments to use special bonds to support rural water supply and sewage treatment facilities
July 2017	Notice on the full implementation of PPP mode for wastewater and waste treatment projects participated by the government	Encourage the full implementation of the PPP mode in the field of sewage and garbage disposal and implement the PPP project in a comprehensive and effective manner

(continued)

Table 2 (continued)

Release time	File name	Main content
August 2017	2017 water pollution remediation work program	Strengthen financial security, increase financial input and actively promote the market-oriented investment and financing mode of government and social capital cooperation
February 2018	Guiding opinions on establishing and improving a long-term mechanism for ecological compensation and protection in the Yangtze River economic belt	Encourage the establishment of diversified investment and financing mechanisms for government, market and private participation, guide social capital to actively participate in ecological environmental protection and stabilize the income and expectations of ecological and environmental PPP projects

Source Documentary of the People's Government of the People's Republic of China

4 Chang'an River PPP Project Case Analysis

4.1 Background

4.1.1 Demand for Improvement of Water Quality in Binhai New District

The originally planned Chang'an New District is located in the south of Chang'an Town. The new district is dedicated to building a marine culture experience city and a multi-industry innovation center. However, according to the historical data of the Dongbao River flooding, there has been upstream mountain flooding in recent years. The flood problem is the main constraint to the economic development of Chang'an Town. The water pollution further limits the development of the green economy in the new district. At the same time, the demand for water pollution control by society and residents is very urgent. The Dongguan Municipal Party Committee and Municipal Government have actively made work arrangements for water control and environmental protection before this. For example, in November 2016, the Municipal Party Committee and Municipal Government held an environmental protection work conference to emphasize the importance of environmental protection work and put forward effective control measures; in January 2017, the use of water pollution control work site will enhance environmental awareness. However, such implementation is not expected to swiftly improve the status quo of water pollution.

The implementation site of the Chang'an River Project is located in the northeastern part of the Binhai New Area. It spans the tidal flat culture pond and links with Mosha River and Maozhou River. The total investment is about 680 million yuan.

Table 3 Documents on PPP mode of water treatment in Guangdong Province, 2015–2017

Release time	File name	Main content
July 2014	"Opinions of the general office of the people's government of Guangdong Province on promoting the development of energy conservation and environmental protection industry"	For the treatment of urban sewage and domestic garbage, expand investment and financing channels, such as supporting energy-saving and environmental protection enterprises to issue bonds, short-term and medium-term notes, encourage and guide social capital and private capital to participate in energy conservation and environmental protection
June 2015	Implementation opinions of the Guangdong Provincial people's government on accelerating the construction of urban infrastructure	Encourage the establishment of multi-participating investment and financing channels and encourage social capital to actively participate in the investment and operation of public welfare urban infrastructure (emphasis on waste water treatment industry)
July 2015	Notice of the Guangdong Provincial department of environmental protection on printing and distributing the water quality comprehensive improvement plan of the Xiaodongjiang River Basin (2015–2020)	Encourage government and social capital cooperation as a PPP mode to raise funds for environmentally friendly infrastructure such as sewage treatment in the Xiaodongjiang River Basin. Encourage large peripheral enterprises to invest more in the comprehensive improvement of the water quality
December 2015	Notice on the implementation plan of the water pollution prevention action plan of Guangdong Province	Encourage the development of sewage, garbage treatment and industrial parks, including government design and equipment, equipment, engineering, commissioning, maintenance and management
December 2016	"Dongguan City environmental pollution prevention and control overall implementation plan (2016–2020)"	Improve government and social capital cooperation mechanisms, further encourage social capital to participate in key areas such as eco-environmental protection, government investment projects can give priority to social capital, or give priority to projects that introduce private capital

(continued)

Table 3 (continued)

Release time	File name	Main content
May 2017	"Guangdong province action plan for South Guangdong water clearance (revised) (2017–2020)"	Encourage the introduction of PPP mode in key sewage treatment, waste disposal and other environmental protection to promote "water-saving clean water is everyone's responsibility," the whole idea of the people involved in water quality improvement work

Source Regulations published by the official website of the Guangdong Provincial People's Government in 2018

The water quality survey results of the Environmental Protection Agency in 2015 showed that the water quality of the Maozhou River Dongguan section was poor, and it has been reduced to the surface water V level. The task of improving water quality is very arduous. After the active management of the local government and all parties, by the end of 2016, the progress of pollution control in the Maozhou River Basin was very smooth. The water quality of the main stream of the Maozhou River was significantly improved compared with the same period of last year, and the main pollutant indicators were reduced, such as ammonia nitrogen and total phosphorus. The decline was 47.7 and 36%. However, the sewage treatment work cannot be relaxed, and it is still the key work to develop the economy of the Binhai New Area. The construction facilities of the Chang'an River Project are an important measure to rectify the water pollution of the Maozhou River Basin and comprehensively purify the water ecology.

4.1.2 The Demand for the Investment and Financing System in the Binhai New Area

According to preliminary estimates, from development to construction (including infrastructure construction costs), more than 18 billion yuan of funds are needed, and the capital needs are enormous. 2016 Chang'an Town Government Work Report shows: In 2015, the disposable fiscal revenue of Chang'an Town increased from 1.31 billion yuan in 2010 to 2.36 billion yuan, an increase of 79.7%. In addition, according to the local government plan, in the "Thirteenth Five-Year Plan" period, Chang'an Town plans to invest another 5 billion yuan to accelerate the construction of the "one network, two plants and three rectifications" project, and the construction of the three phases of the near, medium and long term. Improve the water quality of the Maozhou River Basin and achieve the town-wide water control target by 2020, thereby promoting urban renewal and increasing urban attractiveness. If only local governments raise funds, the investment of more than 20 billion will put a lot of pressure on local finances. Therefore, the former Chang'an New District Management Committee took the lead in setting up Chang'an New Area Holding Company as the financing platform and business entity for the development and construction of the new district, and jointly established a construction platform with Dongshi Group (shared in Chang'an New Area Holding Group, which holds 30% of the shares). Jointly, promote the development of new areas.

Therefore, the pilot project of new urbanization in Dongguan pointed out that it is necessary to establish a PPP-based investment and financing mode in 2017, and gradually improve the budgetary fund management and financial subsidy system to support it and jointly serve the new urbanization. On December 28, 2016, the former Chang'an New District Management Committee, Chang'an New Area Holdings Co., Ltd. and Dongguan Yuanqing Investment Co., Ltd. jointly signed the contract for the PPP project of Chang'an River Project. The PPP project of Chang'an River Project officially landed. At the same time, Dongguan City will also set up a government investment fund to guide private capital to participate in more feasible investment projects, truly diversify investment in environmental governance and lay a good foundation for economic development in the new district.

4.2 Feasibility Analysis of Project Adopting PPP Mode

Dongguan Chang'an River Project is a PPP investment project jointly initiated by the Management Committee and Chang'an New Area Holdings Co., Ltd. and Dongguan Yuanqing Investment Co., Ltd. The Chang'an River Project covers an area of 1525.52 mu, and the construction task is relatively heavy. There are three cores. The contents include the newly excavated 5.2 km river channel, the new east and west pumping stations as well as the new Chang'an River's four sluices. The period of infrastructure construction is four years (2017–2020), and the operation period is tentatively set at 10 years; the total investment is 739 million yuan, and the sources of funds include project company financing and government financial support.

In June 2015, as the first batch of PPP pilot project in Dongguan, Chang'an River Project was linked with Rail Transit Line 1, Shenzhen Outer Ring Express Dongguan Section, Yinbo Innovation Zone Infrastructure Construction and Shimahe Estuary Dongjiang Water Source Protection Phase I Project Together they were included in the Dongguan PPP project library. But so far, only the Chang'an River Project and the infrastructure PPP project of Dongguan Yinbo Cooperation Innovation Zone have been signed. The statistics of the Dongguan Municipal Government and Social Capital Cooperation (PPP) project published on the Web site of the Ministry of Finance and the Social Capital Cooperation Center on January 8, 2018, show that the Chang'an River Project is currently included in the 32 PPP projects of the Dongguan City Project Reserve. One of the three projects under contract signing shows the high level of attention and feasibility of this PPP project.

The Chang'an River Project uses a co-financing cooperation mode between the government and social capital. After the project company has expired, the project company shall hand over the project to the designated party in the contract without compensation. Figure 2 shows the organizational management structure of the

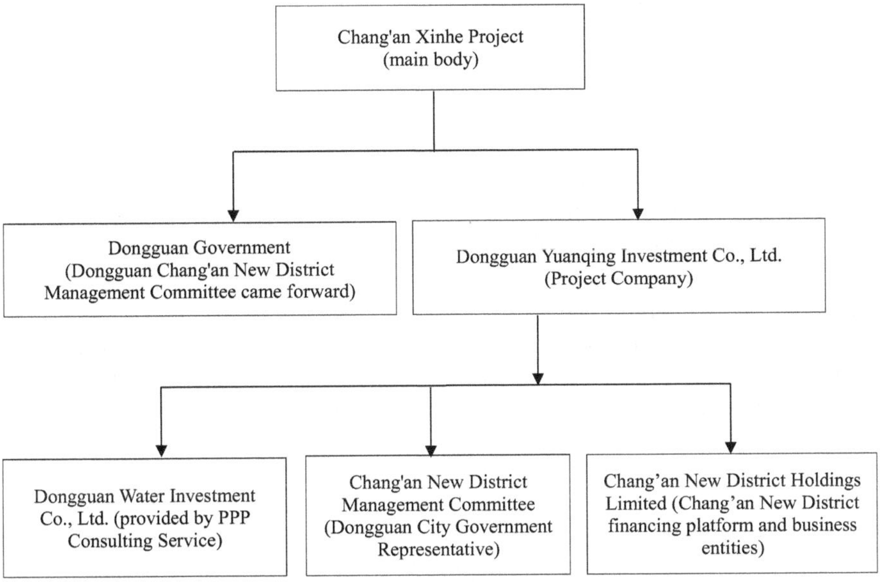

Fig. 2 Organization and Management Structure of Chang'an River Project

public entity involved in the project. Among them, the Dongguan Municipal Government and Dongguan Yuanqing Investment Co., Ltd. are the two main entities of the Chang'an River Project, the essence of which is the project company. The project company was jointly established by Dongguan Yuanqing Investment Co., Ltd. from the former Chang'an New District Management Committee, Chang'an New Area Holdings Co., Ltd. and Dongguan Water Investment Group Co., Ltd., responsible for the financing and construction of the project funds, and under the supervision of the government departments, operate and maintain the Chang'an River, such as repairing facilities.

5 Concluding Remarks

5.1 Challenges

PPP consulting services, technical support and financial support are the main tasks of PPP project operation. Chang'an River Project includes new digging river channel, new pumping station and sluice construction, with a total investment of 739 million yuan. The investment scale is large, and the project is scattered geographically. The technical requirements have become stricter. Moreover, compared with the construction of relatively mature urban water treatment facilities, the project has many problems in the construction and operation stages. The following three aspects are particularly prominent:

First, the Binhai New Area of Chang'an Town, Dongguan City, where the Chang'an River Project is located, is a newly developed construction area in Guangdong Province. Relatively speaking, the government has limited financial resources (higher investment risks in emerging development areas), and its economic strength is weak. There is still a lack of mature technology to support the operation and management of large water conservancy projects. At the same time, due to the poor scale effect of the water treatment project, the total investment is large, the operating cost is high, and it is difficult to attract sufficient social capital.

Second, the above mentioned is that the Chang'an River Project has three major projects that need to be built, including new digging channels, new pumping stations as well as new sluices, which have high requirements for technology. However, the level of urban public facilities construction is different from that of developed cities and regions, especially in the construction of water treatment facilities. The lag of technology would affect the overall operation of the project. Being different from the previous government-funded construction projects, the project can entrust the contractor with more mature technology to carry out the project construction. As the Chang'an River Project has certain public welfare nature and the investment profit is lower than those of the government-funded projects, there is no strong guarantee for the overall construction and operation quality of the Chang'an River Project as a result of such financial constraint.

Third, the design, procurement, construction, commissioning, completion and acceptance of the facilities and equipment of the Chang'an River Project, as well as the initial financial cost estimation of the project, the management of the post-debt service and the post-contract disputes, all require professional technical and knowledge-based talents. The development of Chang'an New District is in its infancy, and the reserve of talent resources is not yet sufficient. In the subsequent development, the cultivation and formation of talent teams will continue. Because the water treatment project has certain special characteristics, it is difficult to carry out scientific management if there is no supporting facilities and technical support. Sewage treatment facilities generally have "three-point construction and seven-point management." Good management would enhance good governance results. If there is no support from professional organizations, management personnel or technical personnel related to water treatment, it would not be able to achieve scientific management effects.

5.2 Implications

First, the effective use of market mechanisms under the PPP mode is to rationally coordinate the interests of all parties. On the one hand, Chang'an River is the first project developed and constructed by Dongguan City using the PPP mode. It will create a precedent for the cooperation of PPP mode in Dongguan and has a strong demonstration and driving role in the construction of other projects. From the above analysis of the financial equity of the project, the Chang'an River Project adopts the BOT financing construction mode to ensure that the market mechanism plays a role, and the social capitalists involved in the investment participate in the preparation and planning in the early stage of the project, which is conducive to integration. And advanced technology and management experience within the diffusion industry.

On the other hand, the government investment company as a social capital party in the early stage of project preparation is involved, so that the project risk responsibility is reasonably distributed in the early stage, which reduces the risks of the contractor and the government, and increases the possibility of successful financing of the project. The strategic alliance of the various participants of the project helps improve project efficiency on the basis of coordinating the different interests of the parties. In the Chang'an River Project, the social capital party has a large operation management right in the project within the time limit stipulated in the contract, and at the same time it is the largest investor. To a certain extent, it can make optimal decisions on project operation management, improvement of operational efficiency as well as increasing revenues generated from the project.

Second, in comparison with other modes of operation such as TOT, DBFO and BT in the PPP mode, the Chang'an River Project still adopts a single BOT method, and the overall operational risk is considered high. Because all facilities of the BOT project must be transferred to the government without compensation, there is a significant risk of cost recovery for the social capital parties involved in such a project. During the operation period, even if it is paid by the government, it is not guaranteed to be absolutely successful in the specific operations. Nevertheless, the pertinent risks are not fully explained in the Chang'an River Project documents. Therefore, regulating the provisions of the PPP contract on circumventing or responding to operational risks is one of the issues that should be noted in the next PPP development in Guangdong Province.

References

1. Abednego MP, Ogunlana SO (2006) Good project governance for proper risk allocation in public–private partnerships in Indonesia. Int J Project Manage 24(7):622–634
2. Fuest V, Haffner SA (2007) Policies, practices and problems in Ghana's urban water supply. Water Policy 9(2)
3. Kim YJ, Choi MJ (2018) Contracting-out public-private partnerships in mega-scale developments: the case of new Songdo City in Korea. Cities 72:43–50
4. Lee N, Schaufelberger JE (2013) Risk management strategies for privatized infrastructure projects: study of the build–operate–transfer approach in East Asia and the Pacific. J Manage Eng 30(3):05014001
5. Mansor NS, Rashid KA (2016) Incomplete contract in private finance initiative (pfi) contracts: causes, implications and strategies. Procedia Soc Behav Sci 222:93–102
6. Moszoro M, Gasiorowski P (2008) Optimal capital structure of public–private partnerships. Soc Sci Electr Publ 08(1)
7. Schuyler House (2016) Responsive regulation for water: balancing commitment and adaptability in the face of uncertainty. Policy Soc 35(2)

8. Shrestha P, Shrestha K, Lidder M (2017) Quality of road maintenance work performed by state force versus private contractors in Nevada. Pract Periodical Struct Des Constr 22(4)
9. Siemiatycki M (2011) Public-Private partnership networks: exploring business-government relationships in United Kingdom transportation projects. Econ Geogr 87(3):309–334
10. Tawalare A, Balu Y (2016) Performance evaluation of implementation of continuous water supply projects: two case studies from India. Procedia Eng 161:190–194
11. Teller J, Kock A (2013) An empirical investigation on how portfolio risk management influences project portfolio success. Int J Project Manage 31(6):817–829

Green Infrastructural Development

Current Trends of Developing Energy Efficiency Projects in the Building Sector of China

Joseph H. L. Chan and Theresa Y. M. Chan

Abstract Energy consumption has been considerably increased in the building sector over the years due to population growth, economic development and increased demand for building functions and indoor environmental quality. Buildings account for over 30% of total energy consumption in China which is one of the world's fastest-growing economies over the globe. China has experienced fast-paced economic growth and has been in a stage of rapid urban development for the last decade; however, challenges from natural resources and sustainable development have accompanied its economic development. In recent years, increasing public concerns on the negative impacts of the construction industry on energy use and climate change has made energy efficiency in buildings becomes a fertile area of research. However, the existing literature consists of diversified topics and issues that when more efficiently classified and integrated for better understanding of energy efficiency in building sector, would pave a way for future scholars and practitioners to undertake more intensive and efficient research. Based on a comprehensive review of contemporary studies on green buildings and energy-efficient buildings, this chapter reviews current issues about energy efficiency in buildings in China and Hong Kong including energy performance contracting, energy efficiency retrofit and net zero carbon buildings, supplemented with case studies in developed cities in China to provide insightful directions for further studies in the area of development of energy efficiency projects.

Keywords Building energy efficiency · Sustainable development · Green building · Energy performance contracting · Energy efficiency retrofit · China

1 Introduction

Green building has attracted worldwide attention both academia and industrial practitioners in recent years [7, 22]. The total carbon emission from the building sector

J. H. L. Chan (✉) · T. Y. M. Chan
School of Professional Education and Executive Development, College of Professional & Continuing Education, The Hong Kong Polytechnic University, Kowloon, Hong Kong SAR, China
e-mail: joseph.chan@cpce-polyu.edu.hk

J. Fu and A. W. Ng (eds.), *Sustainable Energy and Green Finance for a Low-carbon Economy*, https://doi.org/10.1007/978-3-030-35411-4_12

is estimated to be one third of the global greenhouse gas emission [48]. Jiang and Li [18] indicated that the carbon dioxide generated by buildings accounted for around 50% of the total carbon emission which is mainly produced from construction, transportation and industry. China has experienced a rapid urbanisation which has caused considerable pressure on energy consumption, resources and environment. Substantial amount of resources is needed to invest in infrastructure projects and public facilities including housing, transportation and power supply [21]. The emerging environmental issues caused by Chinese urbanisation such as excessive carbon emission and water pollution are highly concerned as China is keeping its pace of urban development. Despite that China's carbon emission is low per capita, it has become the second-largest carbon emission producer behind the United States. The Chinese construction sector has increased carbon emissions and energy consumption, while it is also considered to have high potential to reduce the impact of climate change by developing green building/energy-efficient buildings [48]. As Allouhi et al [2] suggested, improving energy efficiency in building works and civil works becomes more than essential for China and even the globe.

The Guangdong-Hong Kong-Macao Greater Bay Area (GBA) is a China's national key economic development strategy. GBA consists of nine cities in Guangdong Province and two Special Administrative Regions (i.e., Hong Kong and Macao). GBA plays an important role in promoting low carbon and sustainable development. Research about GBA has been emerging in research community. For example, Zhou [53] carried out a CO_2 emission inventories of eleven GBA cities and 12 surrounding cities. The low-carbon roadmap for GBA suggested by the study can provide a benchmark for other countries to adapt climate change and to achieve sustainable development. To et al. [35] reviewed the role of sound and acoustics in the whole life cycle of buildings, in the context of Hong Kong, Macao and GBA. However, not many studies have examined the energy-efficient buildings in the GBA.

Over the first decade of the 21st century, a number of energy-efficient building concepts have been emerged, from energy performance contracting to energy efficiency retrofit and even net zero carbon buildings. As a matter of fact, new buildings have limited impact on overall energy reduction as they represent only a tiny fraction of existing building stocks [43]. Existing buildings can constitute the largest opportunity for energy efficiency improvement. It is therefore important to explore the current trends of energy efficiency initiatives in building sector. Based on a comprehensive review of contemporary studies on green buildings and energy-efficient buildings, this chapter reviews the current issues about energy efficiency in buildings in China and Hong Kong including energy performance contracting, energy efficiency retrofit and net zero carbon buildings, supplemented with case studies in developed cities in China to provide insightful directions for further studies in the area of development of energy efficiency projects.

2 Green Buildings

Public concerns on the impacts of the building industry on energy efficiency and human health as well as climate change have made green buildings a fertile area of research [38]. The terms green buildings, sustainable buildings, high-performance buildings, sustainable construction, green construction and high-performance construction are interchangeable in academic research [15, 16]. Kibert [16] defined green building as a facility which is healthy, designed and constructed with ecological principles and efficiently used resources. Yudelson [47] opined that green building is a high-performance building which has minimal impacts on environment and human health. It is designed to consider and reduce its life cycle environmental impacts as well as water and energy consumption. Sangster [32] considered that the main objectives of green buildings are to (1) minimise energy and other resource utilisation; (2) minimise environmental disturbance and waste generation; (3) boost renewable energy usage and (4) improve health and comfort. In fact, energy efficiency is considered as one of the most significant attributes in assessment of various green building assessment schemes including BEAM Plus, LEED in the United States and BREEAM in the United Kingdom. Ahmad et al [1] pointed out that energy efficiency; reduced maintenance cost and operation cost and extended lifespan of green buildings are the major factors which drive the adoption of green building labelling schemes. It is concluded that green building/energy-efficient buildings present a promising contractual approach for the construction industry to contribute to sustainable development.

Darko and Chan [7] carried out a literature review to examine research trend of green building in ten selected construction management journals. Their study revealed that there had been an increasing green building research interest in recent times, implying that the significance attached to green building by the construction sector was accelerating. The research topics covered in their review focused on green building project delivery, green building certifications, advanced technologies and energy performance in buildings. Buildings account for a significant proportion of the total energy use of the world. The need to reduce energy use has therefore become a key issue in research. They further opined that there is a relatively less research attention paid to issues specifically about improving energy performance of buildings, and it is projected that the interest on this topic will increase and remain to be an interesting area of study in building research. The findings of this study also reinforce the objectives of this book chapter (i.e., to present a review of various current issues of energy-efficient buildings with a comprehensive review of previous research).

3 Benefits and Barriers of Green Buildings

UNEP [37] suggests that the building sector has the largest potential for significantly reducing greenhouse gas emissions compared to other major emitting sectors. It is estimated that the building sector has the potential to make energy savings of 50% or more in 2050, in support of limiting global temperature rises to 2 °C above pre-industrial levels. By the means of fuel switching, energy efficiency and the use of renewable energy, the saving potential of carbon dioxide emission by 2050 in the global building sector can be up to 84 Gt.

Global Buildings Performance Network [10] reports that China emits roughly six billion tonnes of greenhouse gases a year, which is about 20% of total estimated global emissions of 36 billion tonnes. In addition, it was found that 28% of China's carbon dioxide emissions originated from energy use in buildings. It is further suggested that green buildings should be the key to reduce the carbon dioxide emission in China, so as to catch up with the global trend.

3.1 Benefits Associated with Enhanced Competitive Advantages

Developers in China have incorporated green strategies into their business in an attempt to build up branding and reputation and gain competitive advantages at various stages, from planning and design, construction to operation and management. Developers could undoubtedly establish a good corporate image, a valuable intangible asset, whereas green building construction is considered as a commitment on social responsibility. Corporate social responsibility (CSR) is widely perceived as an essential factor affecting a business image and competitiveness. The observation is well supported by studies by the World Wildlife Fund (WWF 2005; [42]) that investing in green buildings can achieve high standards of environmental, social and corporate performances.

The public living environment is greatly improved with the adoption of green roof and provision of high-quality green space. By mitigating the urban heat island effect, the ambient urban temperatures could be greatly reduced. A green branding is always associated with higher price or rental income [50]. Moreover, relationships with governmental planning department, local communities and consumers are also strengthened.

In China, developers can receive lower land prices with the adoption of green strategy. Commitments to the environmental protection in the development are one of the terms and conditions in the contract of land-use rights, and the government in some contexts has been offering developers with favourable land price where green elements and strategies are to be implemented [39]. The government also favours developers with track records of green buildings with tax rebates and financial incentives at the property sale stage as so to encourage development of green buildings.

Increased property values and high rental income are core drivers for green building from the viewpoints of stakeholders. According to the World Green Building Council (WorldGBC) [40], the market value of a property is directly linked to the occupancy and rental rates. Rent levels of LEED-certified buildings outperform traditional office buildings. According to CBRE [4], rental returns are 25–30% higher for LEED-certified buildings than traditional ones. Moreover, the average occupancy rates of LEED and LEED-platinum projects in China was 81.7 and 86.7% which is a 1.5 and 10% higher than traditional offices in 2017.

Office buildings are the main property type seeking LEED certification in China. With reduced utility consumption in green buildings, lower expenses for both tenants and owners are achieved. With higher releasing probability, better-off tenants and higher occupancy rates, more stable rental returns and less costly building operations lead to higher value for green buildings [5]. Increasing number of tenants has displayed concerns for environmental reasons like reduced energy consumption, reduced water consumption, lower greenhouse gas emission and improved indoor air quality, when choosing sites for relocation.

Reduction in construction costs and time from applying green technologies also helps to enhance the competitive advantage of developers. The construction of Vanke City has revealed the implementation of waste reduction procedures and methods help saving considerable construction materials, in a survey conducted by Zhang et al [51]. With sound commitment on time saving, application of prefabrication and integrated elements of construction and decoration are increasingly used in green buildings in China. Moreover, reduction in operation and maintenance costs could also be achieved at post-construction stage with the employment of green technologies like envelope optimisation technology, water-saving appliances and air pressure automatic passive ventilation technology. The market opportunities associated with the development of unique green building product are another potential competitive advantage. With a trend of higher quality living environment expectation, developers could lead the market and gain competitive advantage by establishing a differentiated position in delivering design for comfort and improved indoor air products.

3.2 *Barriers in Developing Green Buildings*

3.2.1 Policy Incompetence

Instead of listing general barriers as illustrated above, this section highlights the core barriers in developing green buildings in China. The development of green building in China is hampered by the insufficiency of green policy implementation efforts. Other than demonstration projects, it is rare for developers to gain subsidies from the Government to develop various kinds of green buildings. Chinese government has drawn up several standards since 2008, such as design standard of domestic green building (JGJ26-95) in 1995 and design standard of public green building (green building 50,189-2005) in 2005, and it is seldom for developers to follow the standards due to inadequate financial sources and high operation cost pressure [41]. Although there is a requirement to obtain a green building design evaluation label before obtaining the construction permit, other green building implementation measures at operational phase are on voluntary basis.

Although the National 12th Five-Year Special Programme for Building Energy Conservation required a minimum energy consumption reduction of 10% for public housings, such national level overall targets have been broken down to provincial and city level [4]. Given the regional economic development is unbalanced, most of northern regions with lower economic development level hinder the large-scale adoption of LEED and thus in achieving the energy consumption reduction target. In accordance with the records of the national Green Buildings Evaluation System (GBES), tier-1 cities, like Beijing, Tianjin, Shanghai, Guangdong, Jiangsu and Zhejiang, have a relatively large number of certified green buildings.

However, provinces and cities are allowed to develop their own building energy efficiency standards [6]. China started to develop its own building energy efficiency standards in the 1990s, and therefore, existing buildings built before the enactments of building standards are usually characterised as having no insulation on the building envelope, poor air-tightness, and equipment and appliances with low energy efficiency.

3.2.2 Cost and Time Concerns

Cost consideration is believed to be one of the crucial economic factors in developing green buildings [12]. Tagaza and Wilson [34] estimated that the capital costs of green building projects range from 1 to 25% higher. Building certification and design complexity are core factors affecting the cost of green building developments. Lengthy approval process and implementation of new green technologies adoption may cause project delay and thus negative economic implications.

3.2.3 Technology Deficiency

China is experiencing a lack of professional, standards, experience and device technology at design stage of green buildings. Without special design capability, unreasonable or unsuitable designs may be resulted with the adoption of standard design by equipment manufacturers without fully considering the adaptability to environment and programme economy [48]. The design process is of vital in the three-star certification of green buildings.

Shortage of experienced workers and related construction standards bring the guarantee of project quality at risk, which may adversely influence the implied benefits at operation and maintenance stage. The technical level and expertise of Chinese domestic green building equipment are relatively less developed. For instance, air conditioning compressors for Chinese domestic heat pumps device manufacturers are mainly from Japanese and Korean enterprises, namely Hitachi, Panasonic, Daikin, LG and Mitsubishi.

3.2.4 Market Immaturity

There is growing evidence that stakeholder demand and awareness have a critical role to play in determining the extent of green building development. The capability of existing manufacturers, suppliers and developers is influenced by market demand. With the absence of effective demand in green building and its associated components markets, the high transformation and material cost may prolong the investment payback period of the developers and thus hinder the supply of the life cycle products of green building. Without embracing the big picture of potential benefits of green building practices, high appliance design costs and limited availability of energy saving material suppliers are common hurdles in engaging developers' involvement in construction waste minimisation [36] and thus green building development.

4 Energy Performance Contracting (EPR)

Even though a number of energy conservation and emission reduction measures have been put forward since the 11th Five-Year Plan in China, the administrative energy conservation measures have gradually lost their effect [30]. With the rapid economic growth of mainland China, the energy consumption has been increasing significantly over the past decades. China has been a net importer of energy since 1992 and the largest energy consumer in the world since 2009 [31]. Energy saving and energy efficiency in China are important from a strategic perspective with regard to energy security and sustainability. EPC is considered as a market-oriented mechanism for promoting energy efficiency has been increasingly recognised by government and private enterprises in China, and its development is highlighted in the 12th Five-Year Plan on strategic emerging industries and energy conservation [33]. The application of EPC in China can be traced back to 1996. EPC is mainly applied to energy-intensive industries such as building industry and transportation [51].

In an EPC, the Energy Service Company (ESCO) sets an energy saving objective with the client. The responsibilities of ESCO include providing an energy saving plan, installing energy-efficient facilities, offering maintenance in the contracting period and ensuring energy saving efficiency. Zhang et al [49] suggested that the ESCO is responsible for all or most of the initial investment in energy-efficient hardware. The client will spend little or no money to utilise the energy-efficient services.

In Asia, various kinds of financial models have been carried out to promote the use of EPC in retrofit projects in existing buildings in Singapore and Taiwan. Among different models in EPC, shared saving model and guaranteed saving model are the most common ones [19]. In a shared saving model, ESCOs sign a financing and performance contract with the client who is usually a building owner. The ESCOs are responsible for design, finance and implementation of the EPC project. ESCOs are also required to verify the energy saving during the contract period. A fixed portion of energy saving is paid to ESCOs over a fixed period. There is "zero risk" to the client [44]. In guaranteed savings model, the ESCO designs and implements but does not finance it, although the ESCOs may facilitate financing in some cases. The ESCOs guarantee that energy savings should be sufficient to cover debt service payment [44]. These two models are compared in academic literature. There are different performance standard in the two models. Guaranteed saving models are linked with energy saving level, while shared saving model is linked with the percentage of energy cost saving [11]. The risk sharing patterns are also different in the two models [33]. The major risks in EPC project include commercial risk, performance risk and credit risk. According to Patari and Sinkkonen [29], ESCOs take the commercial risk, and the client takes the performance risk and the credit risk in guaranteed saving model. In the shared saving model, ESCOs take the commercial risk and credit risk, while the ESCOs take the performance risk [11, 29].

A recent study by Zhang et al [51] reported a national wide and comprehensive questionnaire survey on 248 on-going EPC projects in the building sector in China. The aim of the study is to assess the situation of EPC and to explore if EPC can help China to mobilise potential of energy saving in building sector. The findings suggested that around 90% of the projects were public buildings including government offices, schools, universities and hospitals. Interestingly, EPC, as a market-oriented instrument, should be more favoured in the private building sector. However, it is not the case in China as EPC in building sector is mainly driven by the government energy efficiency policy. In fact, similar findings were revealed in an earlier study by Beerepoot and Beererpoot [3] who investigated the impact of government regulations on innovation and EPC in the building sector in the Netherlands. They concluded that more stringent government regulations can serve as an incentive to innovation, and innovation in the building sector is a process to apply incremental modifications to comply with new and stricter government regulations and standards. The power of government policies was further assessed in Zhang et al [51]. The stricter regulation and compulsory standard were the most powerful policy factors to scale up the application of EPC concept in the Chinese building sector, particularly, energy efficiency for public buildings was compulsory. The owners of public buildings therefore had to find solutions for enhancing energy efficiency, and EPC was considered as one of the most suitable options. Yuen et al. [46] suggested that in order to promote the application of EPC, the Chinese government has composed a three-level policy system including laws, policies and plans. However, there are some issues to be tackled to facilitate a better environment for EPC. They opined that specific laws on EPC, improvement and expansion of scope of EPC could realise the benefits of EPC and hence improve building energy efficiency in China.

4.1 Performance of EPC Projects

There appears to be mixed views of the performance of EPC projects worldwide. A total of eight EPC projects were visited to review the performance of EPC after completion [51]. Interestingly, the energy performance in seven projects decreased by 18.4% on average, as compared with the last year of the EPC contract. The research team [51] opined that limited participation of building owners in EPC period; absence of energy team to manage EPC invested facilities; lack of communication and cooperation between client and ESCO and lack of further investment for maintenance of EPC solutions could be the possible reasons for the failure of these projects. Only one EPC project had made additional investment, and its building energy efficiency increased by 5.5% after completion of the EPC contract. The client was an international owner, and there was a professional energy team involved in this successful case. However, Lee et al [20] proposed a simulation-based method to evaluate the probability of energy saving shortfall of EPC and documented a case study of replacement of heat rejection system for a central chiller plant with three composite buildings in Hong Kong. Their study concluded that the possible energy saving after one year retrofit period ranges from 2.83 to 10.8% with 90% statistical significance. The assessment method can be useful to assess the performance risk encountered in EPC in general.

4.2 *Risks of EPC*

Previous research on risk management of EPC focuses on risk assessment models and risk mitigation measures [19]. For example, risks inherent in performance-based energy efficiency projects are identified and classified into five categories, namely economic risks, contextual risks, technology risks, operation risks and measurement and verification risks [26]. Lee et al [20] proposed the risk assessment model for EPC with Monte Carlo Simulation. A questionnaire survey was conducted with the clients and ESCOs in Hong Kong to solicit their views on risk management of EPC. Their study concluded that payment default risk by client, accuracy of baseline measurement and high cost of installation were considered as the top three key risk factors by the ESCO respondents. Furthermore, the client respondents perceived that complexities in procedures, long payment period and worry about guaranteed saving not being achieved were the three major reasons why many client organisations in Hong Kong not considering EPC implementing in their properties. In order to facilitate wider adoption of EPC, different countries/regions have developed incentive measures in the past decade, including development and promotion of standard form of contract for EPC, ESCO accreditation schemes as well as modification of procurement guidelines. The same study compared the effectiveness of these possible measures and found that "promoting successful examples of EPC projects", "modification of government procurement practices to facilitate the use of EPC contracts" and "government backing up a portion of ESCO's guarantee to lending banks" are top three practical measures for better promotion of EPC practice and achieving higher energy savings.

5 Energy-Efficient Retrofit (EER)

Considering the whole life cycle of a building, operation stage accounts for more than 80% of the energy consumption. Energy-efficient retrofit (EER) for existing buildings has therefore attracted attention internationally, due to its substantial potential on energy saving [49]. Importance of EER has been well recognised by the Chinese government. For instance, 400 million m^2 of residential buildings and 60 million m^2 of commercial buildings were required to be retrofitted as pilot projects to enhance building energy efficiency at national level [28]. Liang et al [23] were of the opinion that most retrofitted public buildings were state-owned including schools, hospitals and government offices, and participation of the private sector was rather limited. Some studies have addressed the insufficient participation of private sector in EER from the perspective of effectiveness of incentive policies. For example, Kasisvanathan et al. [14] indicated that subsidies may not be adequate to cover the long payback period and high cost in EER. A recent research by Liang et al [23] developed a model based on agency theory in order to analyse the decision-making behaviours of various building owners and to optimise the incentive policies in EER, with consideration of building conditions and owner characteristics. The effectiveness of the model was verified by a case study, and it was found that the model was sensitive to different factors including unit cost and energy price. It was further suggested that the incentive policy should be adjusted with the fluctuation of energy price to increase the effectiveness of the incentives.

Hou et al [13] conducted a comparison between the building EER policies in Shanghai, Tianjin, Shenzhen and Chongqing at city level. Their study concluded that the policy design of EER of existing commercial buildings should be further improved, in particular, in terms of subsidy level, number of instalments as well as business model promotion. This study recommended that the Chinese government should craft policies taking into account the local economic conditions, energy saving potential and local commercial building energy efficiency baselines. With regard to technical solutions in Chongqing, lighting system retrofit was the most common retrofit subjects, while building envelope system was the least common one. Yet financial incentives are needed for the upgrades of the building envelope systems. The Chinese building EER market potential is substantial. The coming 10–15 years are a huge opportunity for Chinese commercial building sector to enhance its energy efficiency.

Xu et al [44] explored the success factors for EPC for sustainable building EER of hotel buildings in China. A series of semi-structured interviews and a questionnaire survey were carried out to identify the critical success factors of EPC for building EER projects. With the factor analysis of 21 success factors in the questionnaire survey, six clusters of success factors were identified namely (1) project organisation process, (2) EPC project financing for hotel retrofit, (3) knowledge and innovation of EPC, sustainable development and measurement and verification, (4), implementation of sustainable development strategy, (5) contractual arrangement and (6) external economic environment. The results revealed that the EPC team, client, ESCOs and other related parties who are involved in the EER projects for hotels all can significantly affect the success of the projects [44]. They recommended that education programmes should be provided by government, hotel associations and ESCOs to facilitate knowledge/experience sharing and improve awareness among participants. Furthermore, economic incentives and standard contract procedures and M&V (measurement and verification) protocol can be developed to achieve success in building EER projects in China.

In order to quantify and measure thc performance of building EER projects for hotel buildings, a set of KPIs (key performance indicators) at project level were developed with in-depth interviews with experts and researchers, followed by a questionnaire survey [45]. The result of questionnaire survey was analysed with fuzzy set theory. A total of eight KPIs were identified and selected from twelve performance indicators. They are (1) quality performance; (2) hotel energy management; (3) cost performance; (4) project profitability; (5) energy consumption and resources saving; (6) health and safety; stakeholders' satisfaction and innovation and improvement. By applying the KPIs, the performance of hotel building EER can be assessed objectively.

The market interest in investing in EER projects highly depends on the return of interest. Liu et al [24] conducted a cost-benefit analysis for a retrofit project in a residential building in China in order to assess the economic attractiveness of EER. The study developed a framework to carry out cost-benefit analysis with calculation of cost and benefit over the whole life cycle of existing buildings. The framework was applied in a residential building in China, and it was found that EER generally resulted in satisfactory energy conversation. However, cost effectiveness of various retrofit measures could not meet the expectations. This finding may be due to the low energy price in China. The selectivity analysis indicated that energy price was a key factor affecting the economic viability of building EER.

Zhou et al [52] employed a case study approach to examine the factors affecting the effectiveness of EER in China. The study explored the building EER factors in both design stage and operation stage, where focus was placed on both technology and operational management of building. Zhou et al [52] concluded that conventional building technologies (e.g., improving insulation of building and retrofitting HVAC system) can be applied in the case study, while operation strategies play a key role to realise the benefits of EER. It is not necessary to apply renewable energy technologies with high installation costs and considerable pollution. However, using traditional technologies can help to achieve green building certification if a well-designed building management strategy is in place. To summarise the various studies on EER, it appears that this issue has been studied at national level, city level and project level. As Darko and Chan [7] suggested, EER in building sector is expected to be an interesting and valuable area of construction research in the near future.

6 Net Zero Carbon Buildings

As Feng et al. [9] suggested, the development of green buildings in China began relatively late compared to developed countries. In 2012, the Ministry of Finance and Ministry of Housing and Urban-Rural Development jointly issued "Implementation Guidance on Accelerating in Development of Green Buildings in China" and in 2013, the National Development and Reform Commission and Ministry of Housing and Urban-Rural Development jointly issued "Green Building Action Plan".

A new office tower in the heart of Beijing is the first internationally certified green building in 2005 [17]. Created under the guidance of Natural Resource Defence Council experts and the American-Chinese Coalition Organised for Responsible Development in the 21st Century, the eight-story office tower, with 130,000-square-foot building has used 73% less energy and 60% less water than a conventional office. The building has become a model of sustainable design for builders all over Asia Award won: China's first ever Leadership in Energy and Environmental Design (LEED) gold certification, which is one of the worldwide standards for building sustainability and China's first Green Building Innovation Award.

Green buildings can reduce negative impacts on the environment by using less energy and water. The GM International Operations in Shanghai, which was awarded Gold Certification by the US Green Building Council's Leadership in Energy and Environmental Design (LEED) programme, are a good example in this regard. Up to 90% of its interior space was designed to receive natural light so that the need for ambient lighting is largely reduced. Furthermore, rainwater harvesting system was adopted to collect rainwater from its roof, as an effective means to reduce the water consumption since the rainwater can be used to flush toilets with the combination with wastewater from the hand sinks. As for the benefits of these green features, the building is 16.5% more energy efficient, and it uses 30% less water than standard office buildings in China.

The development of green buildings in China has undergone major advancements in 2017. China has accumulated over 520 million m^2 of green building space at the national standard in 336 cities by 2016. CBRE [4] estimated that, with the inclusion of LEED-certified projects, China currently has over 600 million m^2 of accredited green building space which will be accelerated with the implementation of the 13th Five-Year Plan of Green Building Development, 2016–2020.

Moreover, China is playing a dominating role in promoting zero carbon building in the global platform. With a total of twelve members in the C40, a global network covering over 90 of the world's largest cities, China is determined to convert the Paris Climate Agreement from aspiration to reality. The term "zero carbon", a zero net emissions of carbon dioxide from all energy use, was originated in the Code for Sustainable Home in the UK [8]. The definition of zero carbon is revised in a three-tiered hierarchical approach based on energy efficiency, on-site low or zero carbon technologies and allowable solutions.

The World Green Building Council (WorldGBC) defines net zero building as a building that is highly energy efficient and fully powered from on-site and/or off-site renewable energy sources [37]. Buildings that are energy efficient and supply energy needs from renewable sources are a more appropriate target for the mass scale required to achieve Paris Agreement levels of global emission reductions. C40 is promoting a global movement called the "Net Zero Carbon Buildings Declaration" developed in partnership with the World Green Building Council (WGBC). In order to enrich the Building Energy 2020 Programme, a wide range of energy efficiency policies have been planned and implemented through the C40 China Building Energy Efficiency Programme in the National 13th Five-Year Plan (Table 1).

According to Feng et al. [9], nine building developments were acknowledged as ultra-low energy consumption demonstration projects in Beijing in 2018. It covers one third of the overall ultra-low energy demonstration building target, with a total floor space of approximately 100,000 m^2. The nine demonstration buildings include school, low-income housing, rural residential buildings, commercial and public office buildings and a medical rehabilitation centre.

Furthermore, various building energy efficiency guidance, standards and policies are developed. Ultra-low energy building standards are revisited to ensure quality and consistency from design to construction. Financial incentive policies at district level are enhanced and implemented to guarantee that Beijing's ultra-low energy building demonstration site targets are met. Learning from international experience, officials

Table 1 Building energy efficiency programme in thirteenth five-year plan (MoHURD [27]

Indicator	2015 (%)	2020 (%)	Five-year change
Urban new building energy efficiency improvement			+20%
Urban green buildings percentage in total new building construction	20	50	+30%
Green materials percentage in urban new building construction			+40%
Existing residential building energy efficiency upgrades			+500,000 m^2
Existing commercial building energy efficiency upgrades			+100,000 m^2
Average residential unit heating energy intensity decrease		−15	
Urban level commercial building unit overall energy intensity decrease		−5	
Urban building level renewal energy portion of overall power	4	6	+2%
Energy-efficient residential building in all residential buildings	40	60	+20%
Rural buildings that utilise energy efficiency measures in developed area		10	+10%

from Beijing could better assess the feasibility of mandatory city-wide requirements and roadmaps towards net zero carbon buildings with future policy recommendations across multiple building sectors. Better Buildings Initiative Programme is an example [25] (Fig. 1).

The Changning Districts first plan to strengthen its benchmarking policy and enhance building energy data transparency. As Feng et al. [9] states, the districts develop policies to enlarge the scope of reporting buildings and to pilot further energy data disclosure. The Shanghai government provides subsidies for demonstration buildings that satisfy prescribed retrofit requirements. The district further plans to create pilot projects for clean energy systems such as solar photovoltaic and combined heat power for commercial and public buildings. A pilot nearly zero carbon emission buildings in a low-carbon emission zone is one of the biggest projects in the Better Building China Programme at Changning Districts.

The Better Buildings China Programme is another comprehensive energy efficiency programme for the whole building sector of China. Although the programme is on voluntary basis, stakeholders are encouraged to work together to identify barriers and implement energy efficiency and clean energy solutions in buildings. As

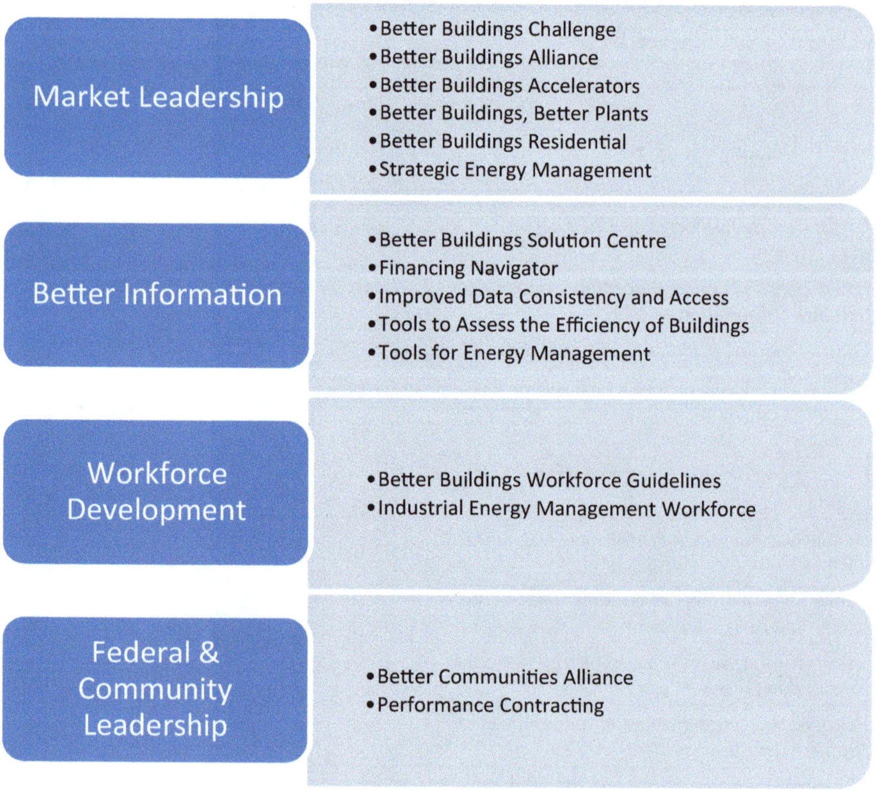

Fig. 1 Better buildings initiative programme structure (Adapted from Liu et al. [25])

Table 2 Better buildings challenge partner progress and results (C40, 2017)

Progress and results	2016	2017
Energy saved (btus)	240 trillion	380 trillion
Dollars saved	$1.9 billion	$3.1 billion
Avoided CO_2 emissions (tons)	15 million	23 million
Funding extended	$8.6 billion	$12 billion
Water savings (gallons)	4 billion	6.3 billion
Partner solutions available online	1000+	1500+
Number of partners and alliance	345+	350+

part of the nationwide programme, the Better Buildings Challenge encourages cities and building portfolio owners to develop energy saving targets and asks participating partners to work together to help achieve the targets of the programme (Table 2).

By connecting China with the global network of more than 90 of the world's greatest cities, urban building policy professionals will be able to align their ambitions with the best practices in promoting net zero carbon buildings, thus delivering the unprecedented pace and scale of the Paris Climate Agreement.

7 Concluding Remarks

Despite its key role in boosting economic development and accelerating the urbanisation in China, the building sector creates environmental problems and consumes a remarkable amount of energy. The huge energy consumption has highlighted the significant need to implement and achieve sustainable development in the building sector and introduction of the green building concept. GBA accounted for 4% of the total national carbon emission and contributed to 13% of Gross Domestic Product (GDP) in China [53]. The potential of nine cities and two Special Administrative Regions (Hong Kong and Macao) will be unleashed, fostering the investment demand and strong economic growth. The recent reforms of the construction industry in Hong Kong, which advocates innovation technology and sustainable development, would deepen the collaboration with the other cities in GBA. Much effort has been put in energy saving in recent years. For example, a memorandum of co-operation (MOC) was signed in November 2018 by relevant institutions and universities in Guangdong, Hong Kong, Macao, Beijing and Shanghai to promote the development and application of retro-commissioning of buildings in GBA for enhancing energy saving. Under this framework, there will be regulator symposiums, technical forums and training courses in GBA for exchange of experience for capacity building of trade and promoting application of retro-commissioning for development of energy-efficient buildings.

With the increasing concerns on sustainable development in recent years in China, energy efficiency in building sector has therefore received more attention from both researchers and industrial practitioners in construction field. This chapter presented an extensive review of contemporary studies on green buildings and building energy efficiency and reviewed a number of current issues about energy-efficient buildings including energy performance contracting, building energy retrofit and net zero carbon buildings in Chinese context, in order to provide useful and promising directions of further research. This chapter can benefit researchers as well as practitioners to synthesise various kinds of information to capture green building research in China. It is hoped that this chapter can enable researchers and industrial practitioners to appreciate the current trends of building energy efficiency in China and expand the existing body of knowledge in this field.

References

1. Ahmad T et al (2016) Developing a green building design approach by selective use of systems and techniques. Archit Eng Des Manage 12:29–50
2. Allouhi A et al (2015) Energy consumption and efficiency in buildings: current status and future trends. J Clean Prod 109:118–130
3. Beerepoot M, Beerepoot N (2007) Government regulation as an impetus for innovation: evidence from energy performance regulation in the dutch residential building sector. Energ Policy 35:4812–4825
4. CBRE (2017) 2017 China green building report from green to health, CB Richard Ellis Group (CBRE) Research. Available at www.cbre.com.hk. Date of access 19 Nov 2018
5. CBRE (2015) The new era of green buildings in China, CB Richard Ellis Group (CBRE). Available at www.cbre.com.hk. Date of access 19 Nov 2018
6. Chan A et al (2017) Barriers affecting the adoption of green building technologies. J Manage Eng. https://doi.org/10.1061/(ASCE)ME.1943-5479.0000507
7. Darko A, Chan APC (2016) Critical analysis of green building research trend in construction journals. Habitat Int 57:53–63
8. DCLG (Department for Communities and Local Government) (2006) Code for sustainable homes: a step-change in sustainable home building practice, p 9
9. Feng W et al (2018) Constructing a new, low-carbon future, How Chinese cities are scaling ambitious building energy-efficiency solutions. In: C40 China building programme launch report, Berkeley Lab
10. Global Buildings Performance Network (2013) Achieving scale in energy-efficient buildings in China. In: A view from the construction and real estate sectors, The Economist Intelligence Unit
11. Han Q et al (2006) Research on commercial patterns of China existing building energy retrofit based on energy performance contracting. In: Proceeding of the international conference for enhanced building operations (ICEBO)
12. Hasan M, Zhang R (2016) Critical barriers and challenges in implementation of green construction in China. J Curr Eng Technol 6(2):435–445
13. Hou J et al (2016) Comparative study of commercial building energy-efficiency retrofit policies in four pilot cities in China. Energ Policy 88:204–215
14. Kasivisvanathan H et al (2012) Fuzzy optimisation for retrofitting a palm oil mill into sustainable palm oil-based integrated biorefinery. Chem Eng J 200–202:694–709
15. Kats GH (2003) The cost and financial benefits of green building: a report to California's Sustainable Building Taskforce, USA. Available at http://www.greenspacebuildings.com/wp-content/uploads/2011/05/Kats-Green-Buildings-Cost.pdf. Date of access 19 Nov 2018
16. Kibert CJ (2007) The next generation of sustainable construction. Build Res Inf 35:595–601

17. Ling Y et al (2015) Developments of green building standards in China. Renew Energ 73:115–122
18. Jiang H, Li JM (2010) Difficulties and strategies of the development of low-carbon buildings of China. China Popul Resour Environ 12:72–75
19. Lee P et al (2015) Risk in energy performance contracting (EPC) projects. Energ Build 92:116–127
20. Lee P et al (2013) Probabilistic risk assessment of the energy saving shortfall in energy performance contracting projects—a case study. Energ Build 66:353–363
21. Li B, Yao R (2009) Urbanisation and its impact on building energy consumption and efficiency in China. Renew Energ 34:1994–1998
22. Li DH et al (2013) Zero energy buildings and sustainable development implications—a review. Energy 554:1–10
23. Liang X et al (2019) Making incentive more effective: an agent-based model for energy-efficiency retrofit in China. Energ Policy 126:177–189
24. Liu Y et al (2018) Cost-benefit analysis for energy efficiency retrofit of existing buildings: a case study in China. J Clean Prod 177:493–506
25. Liu Y et al (2017) An overview of the US better buildings initiative as a model for other countries. Zhou China Energy Group, Lawrence Berkeley National Lab
26. Mill E et al (2006) From volatility to value: analysing and managing financial and performance risk in energy saving projects. Energ Policy 34:188–199
27. Ministry of Housing and Urban-Rural Development (MoHURD) (2017) The 13th five-year plan of China, building energy efficiency and green building
28. Ministry of Housing and Urban-Rural Development (MOHURD) (2012) The 12th five-year-plan of China energy building energy efficiency plan
29. Patari S, Sinkkonen K (2014) Energy service companies and energy performance contracting: is there a need to renew the business model? Insights from a Delphi study. J Cleaner Prod 66:264–271
30. Qian D, Guo J (2014) Research on the energy saving and revenue sharing strategy of ESCOs under the uncertainty of the value of energy performance contracting projects. Energ Policy 73:710–721
31. Qin Q et al (2017) Selection of energy performance contracting business models: a behavioural decision-making approach. Renew Sustain Energ Rev 72:422–433
32. Sangster W (2006) Benchmark study on green buildings: current policies and practices in leading green building nations. Available at http://www3.cec.org/islandora-gb/en/islandora/object/greenbuilding%3A143/datastream/OBJ-EN/view. Date of access 19 Nov 2018
33. Shang T et al (2017) A review of energy performance contracting business models: status and recommendations. Sustain Cities Soc 34:203–210
34. Tagaza E, Wilson S (2004) Green buildings: drivers and barriers: lessons learned from five Melbourne developments. In: Technology report, building commission by University of Melbourne and business outlook evaluation
35. To W et al (2018) Acoustic comfort in buildings in Hong Kong, Macao, and the Greater Bay Area of China. Proc Meet Acoust 35:015001
36. Udawatta N et al (2015) Attitudinal and behavioural approaches to improving waste management on construction projects in Australia: benefits and limitations. Int J Constr Manage 15(2):137–147
37. UNEP (2016) United nationals environment programme frontiers 2016 report: emerging issues of environmental concern. United Nations Environment Programme (UNEP), Nairobi
38. USGBC Research Committee (2008) A national green building research agenda. Available at https://www.usgbc.org/sites/default/files/Docs3402.pdf. Date of access 19 Nov 2018
39. Wang Y et al (1999) Commercial housing development in urban China. Urban Stud 36(9):1475–1494
40. WorldGBC (2013) The business case for green building: a review of the costs and benefits for developers, investors and occupants. Available at http://www.worldgbc.org/files/1513/6608/0674/Business_Case_For_Green_Building_Report_WEB_2013-04-11.pdf. Date of access 19 Nov 2018

41. Wu Y et al (2013) Green building development in China-based on heat pump demonstration projects. Renew Energ 53:211–219
42. WWF (2007) Investing in sustainability, progress and performance among the UK's listed house-builders, WWF-UK, Surrey
43. Xing Y et al (2011) Zero carbon buildings refurbishment—a hierarchical pathway. Renew Sustain Energ Rev 15:3229–3236
44. Xu P et al (2011) Success factors of energy performance contracting (EPC) for sustainable building energy efficiency retrofit (BEER) of hotel buildings in China. Energy 39:7389–7398
45. Xu P et al (2012) Key performance indicators (KPI) for the sustainability of building energy efficiency retrofit (BEER) in hotel buildings in China. Facilities 30:432–448
46. Yuan X et al (2016) Towards a sustainable society: the status and future of energy performance contracting in China. J Clean Prod 112:1608–1618
47. Yudelson J (2008) Green building A to Z: understanding the language of green buildings. New Society Publisher, Canada
48. Zhang L et al (2017) Critical factors of low-carbon building development in China's urban area. J Clean Prod 142:3075–3082
49. Zhang X et al (2015) "Turing green into gold": a framework for energy performance contracting (EPC) in China's real estate industry. J Clean Prod 109:166–173
50. Zhang X et al (2011) Green strategy for gaining competitive advantage in housing development: a China Study. J Clean Prod 19:157–167
51. Zhang M et al (2018) Managing energy efficiency of buildings in China: a survey of energy performance contracting (EPC) in building sector. Energ Policy 114:13–21
52. Zhou Z et al (2016) Achieving energy efficient building via retrofitting of existing buildings: a case study. J Clean Prod 112:3605–3615
53. Zhou Y (2018) Emission and low-carbon development in Guangdong-Hong Kong Macao Greater Bay Area cities and their surroundings. Appl Energ 228:1683–1692

Elevator Energy-Efficient Projects in the Next Generation of High-Rise Green Buildings

Tai Ming Wut

Abstract The chapter investigates the potential energy-efficient solutions for high-rise buildings in urban cities in the Greater Bay Area (GBA) of China. The solutions start from innovative elevator designs that embrace the use of intelligence systems. Smart system maintenance methods are proposed for the existing elevators in old buildings. Total replacement is considered unnecessary since modernization and upgrades would offer a value-for-money option. Energy can be saved through upgrading the control system that enables better comfort on each ride. Such innovation in energy saving from the operations of elevators in high-rise buildings is expected to make substantial contributions to reduction in electricity consumption under the swift urbanization of the GBA.

Keywords Energy efficiency · Internets of things · Elevator industry · Green building

1 Introduction

According to Kone [11], the global market size for new elevator installations in 2016 is around 0.82 million units and that for modernization/maintenance services is about 14 million units. The new installations' markets mainly emerge in China and India while the potentials for modernization/maintenance services are rather fragmented throughout the world [11]. Key players are Kone Corporation (Finland), OTIS (USA), Schindler (Switzerland), Mitsubishi Electric Corporation, Fujitec, Toshiba and Hitachi (Japan), and ThyssenKrupp (Germany).

China has more than 1400 skyscrapers above 150 m of which 48 are particular high (300 m and above). Many of them are situated in the Greater Bay Area. The earliest super tall building in Mainland China is CITIC Plaza in Guangzhou. It was completed around 1996. Ping An Finance Centre in Shenzhen is perhaps the highest one in Greater Bay Area as of today. It has a height of 599 m. It is ranked number

T. M. Wut (✉)
School of Professional Education & Executive Development, College of Professional & Continuing Education, The Hong Kong Polytechnic University, Kowloon, Hong Kong SAR, China
e-mail: edmund.wut@cpce-polyu.edu.hk

J. Fu and A. W. Ng (eds.), *Sustainable Energy and Green Finance for a Low-carbon Economy*, https://doi.org/10.1007/978-3-030-35411-4_13

four in the world. The first runner up goes to Guangzhou CTF Finance Centre with 530 m. Hong Kong's International Commerce Centre only has 484 m [5, 12].

There are various traditional methods to improve the energy efficiency of elevators and do not increase the waiting time of passengers at the same time. They are manual controls and grouping the passengers by zone. Today, Internets of things (IoT) is used to improve further the efficiency. Intelligence system and visualization technology are examples.

Guangdong-Hong Kong-Macao Greater Bay Area composes nine cities including Guangzhou, Hong Kong, and Macao. The total population is over 70 million. It is expected the urbanization rate is over 70% before 2030. There will be over 190 million new urban populations in China as a whole; some of them would be within Greater Bay Area. According to Yang Xianling, President of Ke Research Institute, almost half of new urban population will come to the three urban areas, the Pearl River Delta region is one of them, and the other two are the Yangtze River Delta area and Beijing Tianjin Hebei region [20]. It is expected China would become the largest new equipment, elevator, and escalators in the world. The demand of elevators is becoming greater, especially for high-rise building, in the Greater Bay Area.

Ridder [16] points out in China market, half of new high-rise buildings are offices, and other are residential buildings (20%); hybrid buildings (27%); and hotels (5%).

According to the announcement of the outline development plan for the Guangdong-Hong Kong-Macao Greater Bay Area on February 18, 2019, China aims to develop the Greater Bay Area into a global innovation and technology center. James Yan, Research Director at Counterpoint Technology Market Research, said that "The plan will assist enterprises that are enterprises that are engaged in intelligent manufacturing, high-end telecommunications and mobile internet services in the Greater Bay Area" [6]. In fact, Chinese elevator and escalator companies are gaining market share in their own domestic market recently. One of the reasons is the cost consideration of real-estate developers. They have over half of market share on new installation market and higher percentage on maintenance business.

The objectives of the paper are to introduce how technology and innovation can help in elevator industry and what is the next development in the future. Examples on Greater Bay Area or other cities will be used as illustrations. There are five main parts in the paper: elevator design; Internet of things; maintenance; control system and value-added installations.

2 Innovation in Elevator Design

The process in China requires the manufacturer to develop and verify the design according to ISO 22559 and ISO 14789. The ISO 22259 contains three documents: Global Essential Safety Requirements; Global Safety Parameters and requirements for Global Conformity Assessment Bodies. ISO 14789 Risk Assessment Methodology ensures the proposed design is safe and that all risks have been identified.

Documentation on design and testing records is the requirement [3]. Those documents are then submitted to engineering and testing organizations recognized by the Standardization Administration of the People's Republic of China to evaluate the innovate technology.

There are three types of new features regarding designs in elevator: counter-weightless elevator; double-deck elevator; and emergency rescue system.

2.1 Counter-Weightless Elevator

ThyssenKrupp invented an energy-efficient elevator by using a lightweight car. Transparent polycarbonate panels were used. In order to reduce the weight, the elevator minimizes the use of components, even counter-weight. The drive system consists of a gearless machine and inverter. The power was disconnected when the lift stops. LED lights also shut off automatically which lowers energy consumption further [4]. The restriction of counter-weightless elevator is the limit of serving floors and would be two to three floors.

An electronic monitoring system allows the elevator to communicate with the call center of the lift company directly. The call center receives the signal from the lift showing something wrong. The call center could reset the control system remotely in order to let the lift resume its service as soon as possible. It minimizes the communication between the management and lift companies and increases the response time of rescue team. It is because no mechanic needs in the whole process.

2.2 Double-Deck Elevator

Passenger capacity of the double-deck elevator can be doubled up to 1.8 times compared to traditional single-deck elevator. Environmentally friendly design of the double-deck units includes a permanent magnet motor and power revitalization inverter. Congestion in the lobby area could be reduced by dispensing traffic between the first and second floors. For example, the upper deck serves even floor and lower deck serves odd floor. There are various double-deck elevators installed in Hong Kong. Examples are Hong Kong Island East and Sun Hung Kai center.

2.3 Emergency Rescue System

It is very common in summer, and there are thunderstorms in the Greater Bay Area. In one time, there could be a drop in voltage and caused a lot of elevator breakdowns. There would be many passengers trapped inside the lift cars at the same time. It is difficult for maintenance contractors and firemen to rescue all the trapped people within a short time. Some people, depending on their age and health condition, might feel uncomfortable during the trapped time period.

Shenzhen INVT Electric Co., Ltd is one of the active companies which provides an innovative solution to elevator companies. It is a listed company in Shenzhen Stock Exchange had now has more than three thousand employees. It means company shares could be subscribed by public. An additional instrument like an auto rescue device could be installed in the lift system. When the power supply is interrupted, the passengers were trapped in the elevator. The device can drive the lift car to the nearest floor and open the door safely. Basically, the device turns into normal mode when the power supply is failure. After the desired operation is completed, the device turns off automatically.

Event monitoring device uses the same concept. The device has a sensor to detect the earthquake. In high-rise building, the fireman might not reach the higher floors easily. Thus, an event monitoring device assists the passengers to evacuate. First, the device senses an earthquake event automatically and sends information to elevator controls to take action depending on the magnitude of earthquake. If the event is serious enough, the device signals a controller to stop the car at the nearest floor so that passenger may exit.

2.4 *New Development in Rope System*

A growing height of building, one of the solutions, divides the whole building into different zones so that for those who reach the higher zone could skip lower floors. As the height of building becomes higher and higher, one has to exchange lifts in the middle, like Central Plaza situated in Hong Kong. Can we reach the highest floor without switching elevator? Kone provides an answer. They use carbon fiber material serves as a hoist core, which is covered by a polyurethane coating. This extends the rope's life. For a 500-m tall building (around 130 floors), the annual saving in electricity is up to 15% [14].

2.5 *Machine Room-Less Elevator*

Traditionally, there is always a machine room at the roof so that elevator can move freely through the shaft. Due to the limited space requirement, hydraulic lift system was used so that the machine room can located at the other place. However, there are some drawbacks for the hydraulic lift systems. They are still need a machine room though not necessarily at the top of lift car but somewhere else. The speed of the hydraulic lift is rather slow which is not suitable for more than five stops in general. The use of hydraulic oil would have spill risk and might have a smell during summer.

The design of machine room-less elevator is ideal for add on lift which originally does not have space for the elevator like bridges. No oil design is environmental friendly, and the running speed is faster than hydraulic lift system.

If no machine room at the top, glass could be used instead of laminate plastic walls for the top and more sunlight can be introduced inside the lift car. Less lighting would be used.

3 Internet of Things

According to Visiongain [18], the global Internet of things market will generate revenues of $1128 billion in 2017 and shall be used in different industries including elevator industry [18]. One of the main devices which could be used is the smartphones. Almost everyone has smartphones nowadays, and it is possible to send the signal to the elevator system that you are going to leave the office in 5 min. A lift would be a standby in your floor waiting for your appearance. Of course, if you are late, the lift shall response to the other calls in order to save the waiting time for the other passengers. By the way, you will receive the reminder signal again that asking you your next order time if needed.

3.1 Intelligence System

Objects are connected to each other over the Internet. It could ensure accuracy and efficiency. Human to object pathways can be eliminated or saved. Objects are processed a unique identifier [13]. Physical objects are integrated with Internet and formed a network of related objects [10]. For examples, elevators' positions in a commercial building are detected through an intelligence system so that the control room of building management office could know the exact floor location for every elevator anytime. Management staff is able to reach the floor where one of the elevators stopped and was found idle for a while due to some reasons.

The new development is the management staff's mobile phone which can have lift locations in order to better carry out his or her duties. Management staffs sometimes need to do the patrolling for the whole building. He or she will know the lifts situation anytime in order to have better picture anywhere.

3.2 Maintenance Call System

OTIS Hong Kong uses a sophisticated call system in their maintenance call center in Kowloon Bay. The call center receives emergency calls from their clients everywhere in Hong Kong Area. If there is a trapped person insider the lift car, the emergency team has to reach the site within thirty minutes and then release the trapped person. If there is no trapped person for the case, the emergency team has to reach the site within sixty minutes and then to fix the problem. Otherwise, the management staff will call the Fire Services Department for help. Firemen usually reach the site in a rather short time, say ten to fifteen minutes, depending on the traffic situation. Firemen sometime break the lift door in order to rescue the trapped person. Management Company then blame for the late arrival of elevator company's mechanics according to the signed maintenance contract.

OTIS mechanics carries a Global Positioning System (GPS) device during their emergency repairing; thus, the call center knows their exact position anytime. When the call center receives emergency calls that there is a trapped person, the staff in charge the call center shall decide which team of emergency team will cater for the call. The team chosen may stand by in the station or completing the emergency repair close to the area. By using the said system, it shorted the response time for emergency calls.

3.3 Visualization Using Closed Circuit Television (CCTV)

In traditional response system and destination response system, there is a program in the lifts system to cater for the incoming calls from various buttons on different floors. For example, there are calls on the ground floor for upward direction, tenth floor, sixth floor, twenty floors for downward direction at a particular point of time. The program will respond the calls one by one so that the total waiting time for each "floor" is minimum. One of the drawbacks for the program is to ignore how many people are waiting on each floor. For example, during the morning reporting time, most of the people are waiting on the ground floor. That is to say, this is an upward traffic busy time during before 9 pm. In the past, we count the traffic manually and set the lift go back to the ground floor automatically after responds to the previous call. This is an example of home system.

During the downward traffic busy time, usually after 5 pm. All the lifts might park to higher floor in order to respond the calls in a shorter time. It is because people in the office building start to leave their working place. Thus, their path is from a specific floor to ground floor. Thus, it is no need to park the elevator at the ground floor.

Today CCTV system has been installed in almost all the lift lobby for security or other reasons. In order to addressing the drawback of traditional response system and destination response system, CCTV camera captures how many **number** of people waiting for upward traffic (in the ground floor) and how many **number** of people waiting on each upper floor for downward traffic normally. Thus, optimization (minimization of total waiting time for all passengers as a whole) can easily be reached to by sending lift(s) to the floor which most people are waiting in real-time basis.

4 Smart Maintenance

4.1 Remote Monitoring

Remote monitoring is using the technology of Internet to monitor the operation of the lifts by maintenance center 24 h a day. If there are some irregularities occurred, mechanic(s) will be sent on time to fix the problem. It could minimize any delay to restore normal operation and anticipate the breakdown of the elevators [19].

Some lifts installed at remote area and usually mechanics visit the site twice per month for routine maintenance work in order to satisfy the statutory requirement (depends on the regulation at that place). By using remote monitoring, it could reduce the number of regular visit without affecting maintenance quality. Moreover, when there is an emergency call, the central maintenance team shall use remote monitoring to know the situation of the lift. Suitable mechanic(s) with appropriate replacement parts can be identified and send to the site at the same time. Thus, there is no need to wait for site report from mechanic and replacement parts are prepared according to the report. As a result, elevator service can be resumed in the shortest time and save some maintenance logistic time.

Remote monitoring system can send the information to service mechanic's smartphones directly so as to improve the response time. In the past, the call center receives the information first and then passes to an engineer. The engineer selects the suitable mechanic to response the call. It takes time for the whole process.

The remote monitoring system also sends the breakdown report and total daily traveling time to management office automatically. There is no need for management staff to request the report from lift maintenance office. At the same time, the management staff could submit service request by his or her smartphone using apps instead of making verbal call to the call center of lift maintenance company.

4.2 Smart System Maintenance

Smart system can be installed inside the lift pit. If water accumulates in the elevator pit, the device will generate an audible alarm and trigger a signal. When the system connected to the internet, it will send email notifications or messages to the designated smartphone or computer. It is useful because when a heavy rain suddenly comes, management staff is very busy for many things. It might not be aware with the situation of lift pit which is basically unseen. However, the consequence is great because if there is a water damage inside the lift pit, it takes time and money to exchange damaged parts. The water damage is due to the negligence of the management office staff and the lift owner. Thus, it would be good if the system gives a warning to the management staff when the water level reaches a certain level. The relevant staff could put the lift at the top of the building in order to avoid the water damage. It saves the lift owners' money, and more importantly, to avoid the breakdown time due to repair.

4.3 Machine-to-Machine Functions on Elevators

Preventive maintenance strategy is the goal for elevator maintenance companies. Smart elevator solution (SES) collects elevator operation data and provides alerts on time before something wrong. SES collects information and sends notifications to the remote cloud server. If necessary, technician will be notified as well [8].

Big data can offer equipment type analysis. For example, if we know the real usage based on tracking rides and we shall predict the part replacement schedule. Although elevators are built from various parts, few are expensive and require long installations. Identifying potential problems in advance using sensors can save a lot of money and time compared to the fixing the defected parts.

Preventive maintenance can improve the building owners' satisfaction level and lead to fewer calls to maintenance company. As a result, owners are reluctant to change the service provider and become a loyal customer. The spirit is of the preventive maintenance which is a proactive approach rather than a reactive approach. Breakdown maintenance is a typical example of reactive approach. The contractor waits for the breakdown call and fixes the problem. It creates an unsatisfactory mood among the building owners. It is because the elevator cannot be used for a period of time due to the breakdown. In contrast, parts are changed before its expire lifetime. Less breakdown time will be expected and satisfaction levels of building owners increase. The preventive maintenance approach would not increase the maintenance cost of contractor as less emergency calls expected. Old parts are changed during the routine maintenance time.

4.4 Management of Maintenance Contractor Operations

In the past, elevator maintenance contractors rely on manual method or standalone software to support the business operation. Unfortunately, mistakes occur, because of too much paperwork in the maintenance office. Technicians have lack of information when they are at work at sites and rely on memory. Site information, elevator specification, maintenance records, and management staff contact details are essential information in the maintenance operation. Those details should be assessed by all staff anytime with their smartphone.

In the maintenance service center, service scheduling, routing, and dispatching technicians are the functions most desired by elevator companies. Dedicated software is used to integrate the above functions. Routing optimization using Google maps for technician is valuable tool. It helps to reduce travel time to and from project sites for both preventive maintenance and ad hoc calls. The smart software would discover same or nearby site logic assign the emergency calls to the mechanic. Basically, the software would check the same or nearby site preventive maintenance mechanic when the center received emergency call. It could minimize the travel time and improve the response time for emergency calls. Also, it saves some fuel usage and customer would appreciate the speedy services.

5 Value-Added Installations

5.1 Watching TV Screens Inside the Lift Car

Display screens have become an important part of living environment. In mass transit railway, trains, and office building lobbies, there are display screens. Inside elevator car, there could be display screens installed for management building notice, advertisement, and even financial news. Directories could be one of the candidates appearing in the lift car screens [1].

The second consideration would be the location to install the screen. It could be mounted on the main panel right-hand side as well as the main panel left-hand side. If the screen is installed at the back, it will distract passengers' attention. The height of the screen is depended on the size of the lift car. In general, the height will be higher as the size getting bigger.

Regarding the content, office building passengers might interest in financial news. Hotel passengers might interest in local attractions. Residential passengers might interest in weather forecast. University passengers might interest in academic news. Sport facilities building passenger might interest in sports news.

5.2 Usage of Voice Recognition Inside an Elevator

An elevator normally is provided with buttons for specifying a floor number. There is also Braille-type button which can accommodate the blind people. There is one more innovation nowadays with voice recognition device. It is no need to train the user and at the same time the device can recognize different users. The device also connects to a display of the lift so that the user can know whether their instruction can be recognized [15].

Once the passenger enters into the lift car, the user is asked by the system to state his or her destination through a particular phrase: "Please tell me the floor." English and Chinese could be used as the command languages, and the other languages could be used in the future as well. Once the identification of the command, the suitable signal would go to the central unit and then goes to lift controller unit. The system is able to give visual confirmation through the screen or verbal confirmation.

In the event of interruption of the power supply, the central processor finds the breakdown and triggers the backup speech synthesizer. The passenger will be asked if there is a need to communicate with the emergency staff. It will alleviate the working pressure of management staff to help the people in need.

6 Control System and Energy Efficiency

Every building consumes energy. The taller the building, the more energy it uses. The elevators generally consume around 10% of overall electricity of the whole building. Thus, efficiency must be considered when using the elevators. Most of the energy spent by an elevator is during the standby mode. Around half of the energy has been consumed [7]. It would be helpful for saving energy by the development of parking mode function. Manually shut off some of the lifts after office hours for commercial buildings could be considered. Also, the controller can also program to shut off the lift car fan when the elevator is not in use for a period of time, say five minutes in order to save more energy.

6.1 Reverse Journeys

Reverse journeys are discouraged with control system where there are up and down call buttons. People who travel up get into the car when the lift stops on its up way. People who travel down get into the car when the lift stops on it down way. As a result, no reverse journey is happened ideally.

For high-rise buildings in busy hours, passengers deliberately choose a reverse journey by press both buttons with the wish of a faster car arrival. It ensures the car would not be bypass when the lift coming down again with full car. Thus, reverse journey becomes an obvious choice especially for middle and lower floors. It alleviates the psychological feeling not to be bypass. However, the lifts most likely to be stop twice and time is spending. Thus, an intelligence system could by pass the lower floor in order to save downward travel journey. The system either senses the up and down buttons to be pressed almost the same time or sensing the same fingerprint pressing both the up and down buttons. At that situation, the downward buttons would be canceled when the car responses to upward button call. It saves the traveling time and energy ultimately.

6.2 Hybrid System Control

Destination control system had been used by all major lifts' manufacturers. It could be quite efficiency if there is an almost upward traffic such as in the early morning time. After early morning time, there could be some downward traffic and inter-floor traffic, and the efficiency of destination control system varies. Traditional destination control system is inflexible in terms of call reallocation. The lift allocation is fixed once the lift is assigned. The passenger is directed to a particular lift by using a signal or alarm. However, the lift might be delayed because of inter-calls from floors. Also, if the lift is delayed, there is no way to inform the passengers. It could happen when the late coming passengers could take the earlier arrival lift. At the same time, the early arrival passengers still wait for another incoming lifts. Frustration could occur.

In contrast, hybrid control system would cater the real-time changes. When a landing call is registered, the call is allocated to a car which provides the fastest response time. If the allocated car is delayed in the middle, the call can be allocated to a different lift. The whole process is taking place within the control system [9]. Passenger in waiting is not aware in the process. Thus, frustration can be minimized.

6.3 Modernization and Upgrades

Some old buildings use old technology such as motor generator drive for their lift systems. Upgrade to a variable voltage variable frequency drive in the lift system could save up to 40% of energy consumption depending on the elevator type and size [2].

Within elevator cab, replacing traditional particleboard such as laminate plastic walls and wooden ceilings to new particleboard product without added urea-formaldehyde could improve indoor air quality. Urea-formaldehyde is of the main sources of polluted air. Fan inside the elevator cab could also facilitate the air circulation.

Minor upgrade includes changing the lighting system to light-emitting diode (LED) which could also save energy consumption up to 80% [17]. LED is a type of semiconductor which provides light when electricity is used. It is the most efficient lighting source. LED produces directional light which is suitable used inside the lift car. Operating costs are reduced, and replacement cycles are minimized. The lift span can up to ten years and contain no mercury [2].

There is another problem for the lighting. As heat generates from lift car lighting, heat is trapped inside the lift car if the traffic is not busy. It means that we need to turn on the fan in order to avoid overheating inside the lift car. Turn on the fan needs energy [17]. We might consider when the lift is idle, the lift could return back to the ground floor and keep the lift door open. Usually, there is more ventilation in the ground floor lift lobby, and then the heat generated by the lighting could be removed easily.

Table 1 Potential applications of data analytics

Areas of improvement on sustainability performance	Traditional solutions	Potential applications of data analytics	Characteristics of big data
Maintenance: reducing elevator traffic consumption	Identification by passengers; restrictive floor system; manual control	Visualization using CCTV capturing longitudinal data by seasons, daily traffic activities; and types of building with reference to specific culture and habits of the local users	Volume, velocity, variety, and veracity
Maintenance: reducing cost of manpower	Mechanic routine maintenance	Remote monitoring making use of smart meters to measure lift performance	Volume, velocity, and variety
Maintenance: reducing breakdown rate	Mechanics report the defects in elevator manually	Smart elevator solution system collects elevator operation data	Volume, velocity, and variety

7 Potentials in Utilization of Big Data

Table 1 compares tradition solutions with potential applications of data analytics:

In the above table, traffic analysis involves four characteristics of big data: volume, velocity, variety, and veracity. Energy-saving and electricity consumption analysis involve data volume and velocity.

8 Conclusions

Since almost half of our energy spent on buildings, it is anticipated that building could be a valuable source to save energy. Elevator is one part of the building components and usually takes up considerable amount of electric energy, around 10% of total energy building depending on the traffic [2]. New elevator technologies can save up to a quarter of energy compared to tradition devices. Moreover, elevator can even generate energy when the elevator system slows down by fed back into the building's power grid.

With the increasing population in cities, the demand on elevator will become greater. It is expected the growth in demand of building elevators will be at least 10–15% annually for China's Greater Bay Area market alone in terms of new installed elevators [6]. Innovations and energy-saving design in elevator will continue to be a relevant issue for sustainable development in the Greater Bay Area in the near future.

In this relation, it is important to point out that choosing a thorough maintenance program and employing modernization and upgrade elements can make sure energy saving with more carrying passenger power. Preventive maintenance mode must be used since it keeps elevator operation at the optimal state and increases the life time span of the elevators. While it saves the owners' money in the long run, it upholds a policy for health and safety in building operations under a rapid urbanization agenda among the cities in GBA.

All in all, the owners of the buildings with their management staff should maximize the performance of the lift systems through regular modernization, upgrades, and preventive maintenance. Adopting new technologies would bring about environmental benefits and improving the carbon footprints of the building. The elevator companies and maintenance contractors could generate incomes through value-added services and innovative business models. The owners of the building would be able to lower their operating costs as less electricity is consumed. Even more importantly, the building management would be able to focus their time and efforts on exploring value-added development opportunities to reach a lower breakdown rate. The end users of elevators would enjoy improvement in actual quality service as the elevators become more available and reliable at the same time. These progressive enhancements will ultimately create win-win circumstances for the stakeholders.

References

1. Appelbaum B, Weggeland J (2014) Enhancing cab aesthetics with multimedia displays. Elevator World, Aug, pp 104–106
2. Bailey S (2010) Elevating elevator's energy efficiency and performance. Elevator World, Apr, pp 70–71
3. Bialy L (2015) The global relevance of performance-based codes. Elevator World, Apr, pp 116–120
4. Canseco B (2015) Counterweightless elevator. Elevator World, Jan, pp 50–52
5. Eldridge A (2011) World tallest buildings. Britannica Academic. https://academic.eb.com/levels/collegiate/article/Worlds-Tallest-Buildings/570806. Accessed on 20 Sept 2018
6. Fam F (2019) Plans offers big prospects for Bay Area. China Daily. 20 Feb 2019, p 13
7. Francesco G (2015) Door technology for high-rise applications. Elevator World, Oct, pp 154–164
8. Gotlieb G (2015) Advantages of Implementing M2M and cloud services in retrofitted elevators. Elevator World, June, pp 102–107
9. Halsey, L. (2015) "Hybrid Lift Group Control Systems," *Elevator World*, Jan, 104–114
10. Hsu C, Lin C (2016) An empirical examination of consumer adoption of internet of things services: network externalities and concern for information privacy perspectives. Comput Hum Behav 62:516–527
11. Kone (2018) Elevator and escalator markets. www.kone.com. Accessed on 10 May 2018
12. Li Q, He Y, Wang H, Zhou K, Yan B (2017) Monitoring and time-dependent analysis of vertical deformations of the tallest building in China. Struct Control Health Monit 24:1936
13. Miorandi D, Sicari S, Pellegrini D, Chlamtac I (2012) Internet of things: vision, applications and research challenges. Ad Hoc Netw 10:1497–1516
14. O'Connell P (2015) Straight to the top: innovation in the future world's tallest tower. Elevator World, July, pp 74–75

15. Parageorgiou E, Stratigakis N, Asvestopoulos L (2015) Usage of voice recognition inside an elevator. Elevator World, June, pp 108–111
16. Ridder M (2015) Global urbanization. Elevator World, Oct, pp 56–58
17. Shapiro J (2010) The green cab lighting guide. Elevator World, Apr, pp 74–80
18. Visiongain (2018) Internet of things market report, 2017–2022. www.visiongain.com/report/1770/Internet-of-Things-(IoT)-Market-Report-2017-2022. Accessed on 10 May 2018
19. Wang K, Dai G, Guo L (2016) Intelligent predictive maintenance (IPdM) for elevator service. In: International workshop of advanced manufacturing and automation (IWAMA), conference proceedings, pp 1–6. Atlantic Press
20. Yiran Z (2019) Urbanization to fuel rural revitalization. China Daily, 27 Feb 2019

Sustainable Energy in the Greater Bay Area of China: Lessons from the Cruise Industry

Yui-yip Lau and Christelle Not

Abstract Globalization, technological advancement and economic prosperity concertedly generate a significant demand for cruise. In a context of passenger transport, cruise is one of the key passenger modes to foster the regional economic growth and encourage the social mobility in a community. Recently, the development in Southern China and its emerging Greater Bay Area intends for the revitalization of high-end service innovation and boosting up its tourism development. However, the environmental impacts of the cruise shipping research are seriously overlooked. Cruise shipping activities are well recognized as one of the source of air pollution, but they impact the environment more broadly and the stakeholders often underestimated the associated costs and risks of cruising on the environment. International setting of cruising activity and the lack of policy support are still key barriers for the cruising industry to effectively implement renewable and green-energy strategies and sustainable management in general. In this chapter, we will try to estimate environmental impact of the different resources (i.e. energy, water and waste) used during cruising and explore what are the best practices to develop in order to move towards a greener cruising industries in particular in area of rapid development.

Keywords Energy · Water · Waste · Greater Bay Area · Cruise

1 Introduction

A cruise has been identified as the "transportation of pleasure-seeking travellers on ocean voyages offering one or more glamorous ports of calls" [15, p. 360]. Wild and Dearing [28, pp. 319–320] explained cruise is "any fare paying voyage for leisure onboard a vessel whose primary purpose is the accommodation of guests and not

Y. Lau (✉)
Division of Business and Hospitality Management, College of Professional and Continuing Education, The Hong Kong Polytechnic University, Kowloon, Hong Kong SAR, China
e-mail: yuiyip.lau@cpce-polyu.edu.hk

C. Not
Earth Sciences Department, Swire Institute for Marine Sciences, The University of Hong Kong, Hong Kong SAR, China

J. Fu and A. W. Ng (eds.), *Sustainable Energy and Green Finance for a Low-carbon Economy*, https://doi.org/10.1007/978-3-030-35411-4_14

freight normally to visit a variety of destinations rather than to operate on a set route". In general, cruise ship are like "floating hotels", selecting specific ports to provide their customers with fantastic in-port experience [17]. In order to maintain customer service level, cruise ship operators need to provide sufficient resources such as potable water, food and energy to the cruisers onboard. Due to the size limitation and isolation of cruise ships, cruise activities rely on non-local resources and require outside intervention to supply water, food and energy to cruisers. In addition, cruise ship operators need to manage waste produced onboard. Resource's usage and waste management generate pollution which impacts the environment; however, such impact from the cruise ship industry has been regulated since 1973 by the International Convention for the Prevention of Pollution from Ships [14] (MARPOL). MARPOL Annex IV, V and VI are particularly important for the prevention of the pollution to the marine environment since they regulate the impact of wastewater, solid waste and air emission, respectively. However, the rapid growth of the cruise ship activities over the last two decades poses an increasing pressure on the marine environment, which is still poorly investigated especially in Asia. As detailed above, the Greater Bay Area (i.e. hereafter called GBA), which includes Shenzhen, Guangzhou and Hong Kong cruise ports, representing more than 11.94 million cruisers per year is a good example of the rapid growth of the cruise ship activities in these area. We have summarized the main cruise ports among GBA in Table 1. Despite the importance of the GBA environment (air and water quality, marine resources, etc.) on its technological and economic development, there is no evaluation of the cruise ship activities impacts on the environment. In the next sections, we will review the impact of cruise ship industry on the GBA by looking at the management of water, waste and energy done by ship sailing within the GBA. It is worth to note that the management of water and waste is closely linked to the need of energy; therefore, the use of water and management of waste produced onboard is closely tied up to the energy demand, and consequently, the development of sustainable energy use within the cruising industry will impact the global environmental footprint of cruisers.

Within GBA, there are four main cruise ports that are now located in Hong Kong (i.e. Ocean Terminal and Kai Tak Cruise Terminal), Shenzhen (i.e. Shekou) and Guangzhou (i.e. Nansha). Ocean Terminal was a first-class professional cruise terminal for liners and the first large-scale commercial shopping mall in Hong Kong. Due to a capacity limitation, HKSAR government decided to establish a new cruise terminal at Kai Tak to accommodate mega cruise ships and receive the increasing number of cruise liners calling during the peak season. However, a lack of transport connectivity and construction defects pose a problem to attracting cruise liners and cruisers from selecting a cruise terminal at Kai Tak in the final choice. Although Kai Tak cruise terminal strives to carry out comprehensive promotional campaign and progressively improve cruise terminal operations, the capacity underutilization problem still exists [17]. Regarding to Guangzhou, the Chinese government declared *Guangzhou Nansha New District City Master Plan* (*2012-2025*) to develop a large scale of cruise tourism through building Nansha new district, creating yachts products and upgrading cruise port facilities, and integrated with various tourism and

Table 1 Summary of cruise ports and their activity in GBA

City	Number of cruise ports	Number of cruise ship/year	Number of cruisers/year	Companies of the cruise ship that go to that cruise port
Guangzhou	1	104	401,100	Costa, Star Cruises, Dream Cruises
Shenzhen	1	78	188,500	Royal Caribbean, Costa, MSC Cruises, Star Cruises, Silversea, Skysea
Hong Kong	2	249	604,300	Azamara, Costa, Crystal, Cunard, Fred. Olsen, Hapag-Lloyd, Holland America, Norwegian, Oceania and P&O, Princess, Regent Seven Seas, Royal Caribbean, Seabourn, Silversea, Star Cruise

Source Cruise Lines International Association [6], Pinchain [21], Ocean Terminal [20]

infrastructure projects within the Pearl River Delta. For Shenzhen, the Chinese government organized Shenzhen Cruise Carnival with different cruise companies and travel agencies in 2017. Different cruise tourism promotion activities are jointly conducted by travel agency, industry practitioners and cruise liners to promote the cruise tourism widely in the community [21].

The rise of middle class in China stimulates the growth of the cruise industry. Because of geographical advantages, GBA contains coastline cities to boost up the cruise economy. As a result, the numbers of cruisers and cruise ships have been increased dramatically in the past few decades [13]. In the forthcoming years, we expect cruise ports in GBA will progressively increase and occupy a key position in the Asia region. As CLIA (2018) indicated that the Hong Kong cruise passenger CAGR is 20%, we expect that Hong Kong cruise passengers will be reaching at 725,160 and 870,190 passengers in 2019 and 2020, respectively. In Shenzhen and Guangzhou, the cruise passenger CAGR is 29%; we predict that cruise passengers in Shenzhen will be reaching at 243,165 and 313,683 passengers in 2019 and 2020, respectively, while cruise passengers in Guangzhou will be reaching at 517,419 passengers in 2019 and 667,471 passengers in 2020. Sun et al. [26] indicated that cruise tourism produces various supporting industries like hospitality, exhibition, logistics, finance, education and training, marine insurance, ship registration, ship repair and maintenance, and ship chartering. Since the cruise sector and its supporting industries are categorized as a high-end service, it can encourage the transformation and upgrading of the GBA regional economy and finance centre. However, impacts of cruising activities on the environment can also slow down the GBA rapid economic development.

2 Water Management

Cruise ship has a high demand on water to fulfil consumption, personal hygiene, food preparation, cleaning and recreation needs. Therefore, a large amount of good-quality water is needed onboard. Water is either loaded from land and store within the ship or produced via desalination plants. To ensure the quality of the water, several treatments such as filtration, disinfection, mineralization and softening are required, which increase the energy demand and its impact on the environment (see energy section). If the water supply has not a big impact on the environment apart from the energy consumption linked to its treatment, wastewaters, which are the waters collected after their use, are much more an environmental concern. Depending on their usage, wastewater will contain different types of substances such as soap residue, toxic chemical, bacteria, virus, organic matter, etc., which are potentially harmful for the marine environment and therefore require special attention.

Wastewaters are mostly separated between grey water and black water or sewage water. Black waters are produced in much lower volume than grey water but contain higher level of pollutant. Black water or sewage water is the wastewater from the use of toilet and medical facilities onboard the cruise ship. An average of 30 L of sewage per person and per day is produced onboard (EPA 2008a:2-1 see Klein [16]). The presence of harmful bacteria, pathogens, viruses and organic matter content within sewage water requires their treatment before any potential release into the environment to limit their impact. However, MARPOL Annex IV require the treatment of sewage water before their release within 4 nautical miles (7.4 km) of the coastline; outside this area, untreated sewage water is allowed. However, company members of the International Council of Cruise Lines (ICCL), which represent 16 major cruise lines, only allow treated sewage water to be discharged in high sea. Since the cruise ships sailing within GBA are also part of ICCL, no sewage water should be discharged within GBA waters.

Treatment of black water from cruise ship is mostly done by a type II marine sanitation device (MSD), which consists of a flow-through system that breaks up and disinfects (chemically or biologically) waste before discharge. More recent ships have been equipped with advanced wastewater treatment system (AWTS) or even more recently with advanced wastewater purification systems (AWPPS). Most MSD type II and AWTS system filter solids from sewage water as part of the treatment. However, neither MSD nor AWTS do adequately address nutrients loading as nitrogen and phosphorus [16]. Generally, sewage sludge from AWTS can be dumped into the sea beyond the 4 miles' limit; however, its decomposition will induce high oxygen consumption, affecting marine life. Sewage sludge can also be dewatered and then incinerated onboard, which will affect air quality emission from the ship or stored onboard for treatment in port. AWPPS are used to treat black water and grey water before their discharge into the ocean by producing a clean effluent, with limited nutrient concentration, that can be discharged within the 4 nautical miles' limit [16, 27]. In GBA, there are different action plans for tackling water-related pollution problems including the establishment of adequate sewage infrastructure to serve the planned facilities in Kai Tak cruise terminal development [11]; creating Harbour Area Treatment Scheme surrounding Ocean Terminal [10]; adopting advanced waste treatment technology and optimizing waste treatment system in major China cruise terminals [12]. Basically, cruise ports usually grant permit to some cruise liners providing reception facilities to handle wastewater and sewage. To a certain extent, most of the cruise ship sailing within the GBA mainly uses cruise port facilities.

Grey water represents the wastewater from sinks, showers, laundry and cleaning activity of the ship. An average of 120–300 L of grey water per person and per day is produced onboard cruise ship. Even if grey water may contain similar harmful component for the environment as black water, MARPOL does not restrict their discharge in coastal water [3]. GBA also faces the same situation. However, EU and some countries have set restrictions for grey water discharge within their territorial water. To overcome the shortcomings, GBA plans to boost environmental collaboration to reduce emissions, achieve resource efficiency, reduce waste and protect biodiversity [24]. For further protection of the marine environment and nature conservation in the region, Guangdong and Hong Kong signed the 2016–2020 Cooperation Agreement between Guangdong and Hong Kong on Environmental Protection in September 2016. In 2016, Hong Kong and Guangdong established the Hong Kong-Guangdong Marine Environmental Management Special Panel to strengthen communication on marine environmental issues (Great Bay Area, 2018).

Finally, oily bilge water waste is also produced in a significant amount onboard cruise ship. This wastewater contains fuel, oil and wastewater from the engine. Based on MARPOL Annex I regulation, oily bilge water waste can be discharged within the 12 nautical miles, if oil content is less than 15 ppm. Collection and treatment of oil wastewater exist in most of the important port; however, they are located within port facilities and not at the cruising terminal, making the collection of such wastewater during cruise difficult. However, more and more cruise liners perform environmental responsibility to install cruise ship bilge water treatment to remove oil leaking from machinery in accordance with all international, regional and national laws and regulations covered by the GBA [22].

The presence of harmful bacteria, virus or pathogens within wastewater and their release into the environment can be easily recognized, and its impact is well recognized by the impact on marine organisms or humans if the release happens close to the coast. However, the release of high amount of organic matter does not have a direct impact on human health; however, the high demand in oxygen linked to the decomposition of the organic matter is also an important threat to the marine environment. In fact, the high demand in oxygen will lead to the decrease of the oxygen present in the water and the formation of hypoxic to anoxic water, which cannot sustain marine life. Within the GBA, the high nutrient concentration release by the Pearl River itself and the sewage water from the surrounding cities already affect the oxygen level of water of the GBA. Two zones of hypoxia are presently observed seasonally within the PRGBA [1, 23, 25]. The expansion of the cruise ship industry and the potential release of higher concentration of nutrient link to wastewater will potentially further increase spatial extend and/or duration of hypoxia event, leading to more pronounced impact of the marine environment. On top of the impact of low oxygen content on marine life, large and long hypoxia period will also affect human life quality by the presence of smelly water. Such impact could also affect the economy of the region by creating a less attractive area for innovation and technology development.

3 Source of Waste and Their Management

Cruise ship has been estimated to produce almost 25% of the waste from merchant fleet even if cruise ship represents only 1% of the fleet [9]. The high number of passengers is responsible for this overrepresentation of the cruise ship activities. Herz and Davis [9] estimated that at the end of 90s, each passenger produces an average of 3.5 kg of solid waste per day. However, active waste reduction practices in the recent years allowed the decrease by more than half of the solid waste production per person onboard cruise ships [16]. In addition to wastewater, a large amount of waste such as plastic, paper, cardboard, food waste, cans and glass are produced by passengers of cruise ships.

Historically, waste produced onboard ships were disposed over the side of the vessel while at sea. Waste management from ship started in the late 60s, when the potential impact on the marine environment of the indiscriminate dumping into the ocean was first recognized. Under MARPOL Annex V, no solid waste in any form can be released in coastal water within three nautical miles of the coastline. Grounded solid waste, capable to pass through a 2.5-cm screen, can be discharged in water between 3 and 12 nautical miles, and most solid waste can be released into the water beyond the 12 nautical miles' limit. Plastic waste is one of the exceptions, and there is a complete ban on their release into the marine environment, and certain types of plastic are also prohibited to be incinerated.

The management of waste produced onboard cruise ships is linked to onboard facilities (i.e. presence of incinerator) and management plans and facilities at home ports and ports of call. Different methods are used to reduce and management solid waste onboard cruise ship (see Table 2). Recyclable items such as glass and aluminium are increasingly separated and stored for onshore disposal and recycling, whereas the remaining waste, which represent 75–85%, are incinerated. Ashes produced by the incineration can be discharged at sea following MARPOL Annex V guidelines [16]. It is worth to mention that the use of incinerators onboard are far less regulated than that the one on land. MARPOL Annex VI only bans incineration of certain harmful substance from contaminated plastic packaging and polychlorinated

Table 2 Possible shipboard waste treatment processes

Treatment	Method	Management	Type of waste
Compactor	Reduce volume by compaction	Storage then discard at port	Plastic
Commuters	Shred to pieces smaller than 2.5 cm	Discharged at sea	Food
Pulpers	Reduces paper and cardboard to papier mache	Discharged at sea	Paper/cardboard
Shredders	Use to grind	Stored or discharged at sea	Bone, metal, glass
Incinerators	Burn	Ashes are discarded at sea or store for discharge at port	Material than cannot be recycled under MARPOL Annex V

Source Butt [3]

biphenyls (PCBs), and no national or international regulations limit the emission from ship incinerators. New waste management facilities such as Plasma Arc Waste Destruction Systems (PAWDS), which allow for the elimination of paper, cardboard, plastic, textiles and food waste, are potentially accessible but require huge capital investment and operating cost like human resources management, repair and maintenance, information systems and administration. Cruise liners strive to carry out waste management best practices including invest in design and implement comprehensive waste minimization processes and procedures, environmental training and sustainable wastewater operations [7]. However, the sophisticated cruise ship structure and increasing sizes of cruise vessels impose important questions regarding the waste management. In order to bringing cruisers towards a multifaceted recreational shipboard experience, cruise liners provide a wide variety of onboard facilities consisting of adventurous souls, casino, shops, buffet restaurant, theatre, fun activities, entertainment, spa and fitness, just to name a few [17], which add to the daily-life activities related to waste production.

Finally, the different hazardous waste produced onboard cruise ships such as dry cleaning waste, used paint, solvents, expired pharmaceuticals, batteries, etc., are stored and discarded on port following the port country legislation.

The impact of solid waste is multiple depending on their type and management. Waste brings direct biological impacts influenced by ingestion and entanglements. Also, waste restrained coastal marine life and accumulated on the seafloor. Besides, waste produces negative economic influences in case of washing up on beaches, destroying the aesthetical qualities and lowers tourism revenues [19]. For example, the discharge of grounded solid waste such as glass can potentially physically harm marine organisms, but the size of the grounded pieces limits such threat. However, grounded food waste contains organic matter which once in the water will be decomposed by marine organisms. This decomposition requires the consumption of high level of oxygen, which will affect the availability of oxygen in the water. As described above, for wastewater, the release of material with high content of organic matter within the PRGBA will enhance the hypoxic zone already present, and therefore, such release needs to be monitored to avoid severe consequence on the marine environment. The incineration of these solid wastes as paper or food waste appears then as a good alternative, as it allows a rapid reduction of the volume of waste that needs to be stored onboard. The bottom ashes are the results of the incineration of the solid waste. They are composed mostly of mineral fraction and metals, which are non-reactive within the marine environment. However, the incineration will release several components in the air such as carbon dioxide, nitrogen oxides and particulate matter (PM10 and PM2.5), which are responsible for low air quality. Hazardous waste and plastic cannot be incinerated, and special management is required; therefore, such waste are usually stored onboard until discharged at port to be properly managed within land waste treatment plant of the local facilities at home ports and ports of call. Due to plastic ubiquitous usage in human daily life, plastic wastes are numerous onboard cruise ships. Based on MARPOL Annex V, plastic are prohibited from being released into the sea and certain types of plastic are also prohibited from being incinerated, meaning that plastic waste need to be stored and are further treated following the waste management practice of the city home port of each cruise ship.

4 Energy Needs and Its Environmental Impact

Fuel consumption for engine is the main need of energy for cruise ship. In addition to engine energy consumption, cruise ships have a need of auxiliary energy to support light, ventilation, etc., which is produced by onboard generator. [2] identify that 46% of the energy consumed by a cruise ship is linked to its propulsion, 27% is linked to heat and 27% lined to electric power generation. As ships use a "low-quality" fuel (remains from the crude oil after gasoline and distillate fuel oils are extracted), which contain very high sulphur content, for engines and incinerators, they emit air pollutants such as GHG and sulphur and nitrous oxides. Air pollution linked to cruise ships or shipping activities in general is recognized since early 2000s and have been one of the first environmental impacts tackled by the shipping industry and government. There are two types of pollutants emitted by the consumption of fuel. The ones that impact climate are greenhouse gases (CO_2, CO, NOx or HC), and the ones that have an impact on human health are SO_2, PM10 and PM2.5. International regulations are looking at both types of pollutants. For example, the International Maritime Organization (IMO) has officially adopted a strategy to reduce the greenhouse gas emissions from shipping by 50% by 2050 in comparison to the 2008 levels. In addition, MARPOL Annex VI is regularly decreasing the level of sulphur allowed in fuel used by ships. Until 2012, 4.5% sulphur content was allowed in most of the area, except within the Emission Control Areas (ECAs), where only 1% sulphur was accepted and in EU ports where the sulphur content needs to be lower than 0.1%. In 2020, the IMO will enforce a maximum of 0.5% sulphur content in fuel, whereas ECAs and EU ports will continue with sulphur content lower than 0.1%.

The Chinese government designated a series of Domestic Emission Control Areas (DECAs) in its three busiest port regions (Yangtze River Delta, Pearl River Delta and Bohai Bay) as part of a plan to phase in a low-sulphur marine fuel mandate. Starting in 2017, all ships calling at the three regions will be required to use fuels containing no more than 0.5% sulphur when berthing in port. In 2019, they will have to use this cleaner fuel anywhere within 12 nautical miles of coastlines within the DECAs. Since early 2019, Shenzhen ports and all the ports within the Pearl River Delta require the use of fuel with less than 0.5% sulphur content within the DECA. Hong Kong ports follow the same regulation since 2015. [4] estimated that cruise ships are responsible for 2%, 2.3% and 2.2% of sulphur dioxide, nitrous oxide and PM 2.5 emissions, respectively, of the total emission from shipping activities. We therefore estimated that PRGBA cruising activities emitted, respectively, 3441 t per year, 6182 t per year and 196 t per year of sulphur dioxide, nitrous oxide and PM 2.5, using emissions data from 2007 to 2010. Despite the recent changes in the regulation of the greenhouse gas emission from shipping detailed above, the increase of the cruising activities (number of ship cruise and cruisers, which visit PRGBA's ports) is responsible for the augmentation of cruising-related greenhouse gas emissions in the PRGBA.

The extension of the use of low sulphur content fuel to larger area is largely disapproved by shipping industry due to the cost of such low sulphur content fuel. It has been estimated that it will add 15 to 20 USD per passenger per day to use only low sulphur content fuel (Seatrade Insider 2009). Another alternative to reduce the emissions of air pollutant is to reduce the speed limits as cruise ships approach ports. Several ports in California used this approach in 2009, such as San Diego, Los Angeles and Long Beach have shown a significant reduction in emission of nitrous and sulphur oxide, particulate matter and CO_2 (California EPA, 2009).

Some cruise ships are now trying to use gas turbines as greener alternatives to reduce air pollutant emissions; however, such turbines will reduce the emissions of sulphur and nitrous oxide but produce significantly higher volume of CO_2 [16]. An alternative of air pollutant emission linked to fuel consumption is the Alternative Maritime Power (AMP), which allows ship to use energy produced on land and stored by port authorities. However, AMP is still not in application in China since ports are not allowed to sell electricity; in addition, there is another legal aspect linked to invoices, tax for berthed ship in China port as well as safety issues linked to AMP [5]. In addition, cruise liners (e.g. Royal Caribbean, Costa Cruises and Carnival) develop clean energy and nuclear power to power the cruise ships in the forthcoming years [12].

5 Conclusion

The cruising industry is developing rapidly around the world. An increase of more than 20% of cruise passengers over the last 5 years worldwide has been observed, and 27 new ships have been scheduled to be deployed in 2018, resulting in the presence of 476 cruise ships sailing mostly in coastal water. The PRGBA is not an exception, and a rapid increase in cruise ship and cruise passengers is observed. The cruise tourism generates a large amount of revenues from (1) spending of passengers and crew members on land; (2) fees from other services that the ship has been offered by different parties in and out; (3) fees for repair and maintenance of the waterway; (4) fees for services for the boat recognized at port where the boat docked [18]. As a result, cruise industry boosts up regional economic development, social mobility and job employment.

By its nature, cruise ship activities are not sustainable since they required the external input of most of the resources needed for cruisers to have a multifaceted recreational shipboard experience; therefore, the environmental cost needs to be evaluated in order to fully understand the benefit and impact of cruise ship activities within a specific area. Water, waste and energy are the three "resources" that need to be evaluated in order to estimate the environmental impact of the cruise ships. In fact, both water and waste management require energy; therefore, energy is a key resource to examine. In general, precise evaluation of the production and treatment of water or waste onboard cruise ships is limited. Energy production is a little bit more accessible, but to fully assess the impact of cruise ships on air quality, it is necessary to measure and characterize the emission from shipboard engine and incinerators of several ships directly.

Waste management and emissions from cruise ships are regulated by MARPOL Annex IV, V and VI, which aim to limit the discharge of waste and gas emission within few nautical miles' limit from the coast. However, regulations are still quite permissive but most importantly often violated leading to a non-negligible impact of cruise ship on coastal marine environment. Finally, beyond the few miles' limit, there is barely any regulatory enforcement on waste discharge or gas emissions.

The GBA aims to mitigate environmental pollution, support environmental technology and green buildings, advocate energy saving, waste reduction, and encourage environmental education to establish a sustainable future for GBA [8]. However, until now there is no direct information on the environmental impact of cruise ship industry with the area, knowing the exponential growth of this sector, it is now clear that its environmental impact needs to be investigated in order to develop the best practice which will benefit the economic growth and sustainability of the development of the GBA. In the future, we will conduct an in-depth interview with various stakeholders' attitude towards sustainable energy in GBA and contribute the policy paper in the future.

References

1. Archana A, Thibodeau B, Geeraert N, Xu MN, Kao SJ, Baker DM (2018) Nitrogen sources and cycling revealed by dual isotopes of nitrate in a complex urbanized environment. Water Res 142:459–470. https://doi.org/10.1016/j.watres.2018.06.004
2. Baldi F, Ahlgren F, Nguyen T, Thern M, Andersson K (2018) Energy and Energy analysis of a cruise ship †, pp 1–41. https://doi.org/10.3390/en11102508
3. Butt N (2007) The impact of cruise ship generated waste on home ports and ports of call: a study of Southampton. Mar Policy 31:591–598. https://doi.org/10.1016/j.marpol.2007.03.002
4. Chen D, Zhao Y, Nelson P, Li Y, Wang X (2016) Estimating ship emissions based on AIS data for port of Tianjin. China Atmos Environ 145:10–18. https://doi.org/10.1016/j.atmosenv.2016.08.086
5. Chen J, Zheng T, Garg A, Xu L, Li S, Fei Y (2019) Alternative maritime power application as a green port strategy : barriers in China. J Clean Prod 213:825–837. https://doi.org/10.1016/j.jclepro.2018.12.177clia. Available at https://cruising.org/about-the-industry/policy-priorities/clia-oceangoing-cruise-line-policies/environmental-protection. Access on 10 Feb 2019
6. Cruise Lines International Association (2018) Asia Cruise Trends. Singapore
7. Cruise Lines International Association Inc (2019) Cruise Trends & Industry Outlook. Washington, USA
8. Greater Bay Area (2018) Available at https://www.bayarea.gov.hk/en/home/index.html. Access on 10 Feb 2019
9. Herz M, Davis J (2002) Cruise control. A report on how cruise ships affect the marine environment on behalf of The Ocean Conservancy, 2002
10. HKSAR (2005) Wastewater treatment. Available at https://www.dsd.gov.hk/Documents/AnnualReports/0506/eng/ch01_right.htm. Access on 10 Feb 2019
11. HKSAR (2006) Sewerge and sewage treatment implications. In: Kai Tak development—environmental impact assessment report, agreement no. CE 35/2006(CE). Hong Kong
12. Hong W (2018) Report on the development of cruise industry in China. Springer, Singapore
13. Hung K, Wang S, Guillet BD, Liu Z (forthcoming) An overview of cruise tourism research through comparison of cruise studies published in English and Chinese. Int J Hospitality Manage
14. International Convention for the Prevention of Pollution from Ships (2019) Available at http://www.imo.org/en/About/Conventions/ListOfConventions/Pages/International-Convention-for-the-Prevention-of-Pollution-from-Ships-(MARPOL).aspx. Access on 18 June 2019
15. Kendail L (1986) The business of shipping. Cornell Maritime Press, Centreville, MD
16. Klein RA (2010) The cruise sector and its environmental impact. Bridg Tour Theory Pract. https://doi.org/10.1108/s2042-1443(2010)0000003009
17. Lau YY, Tam KC, Ng AKY, Pallis AA (2014) Cruise terminals site selection process: an institutional analysis of the Kai Tak cruise terminal in Hong Kong. Res Transp Bus Manage 13:16–23
18. Lau YY, Yip TL (2018) The Asia cruise tourism industry: current trend and future outlook. In: 11th international conference of Asian shipping and logistics. Incheon, Korea
19. Mohammed J, Torres R, Obenshain E (1998) Waste reduction at sea: pollution prevention strategies on Miami-based cruise lines. Pollution prevention in the coastal zone. National Pollution Prevention Center for Higher Education, University of Michigan
20. Ocean Terminal (2019) Available athttp://www.oceanterminal.com.hk/en/. Access on 4 Feb 2019
21. Pinchain (2018) 2017–2018 China cruise port development annual study. Available at www.pinchain.com/article/149672. Access on 4 Feb 2019
22. Princess Cruises (2019) Available at https://www.princess.com/aboutus/environmental-responsibility/. Access on 10 Feb 2019

23. Qian W, Gan J, Liu J, He B, Lu Z, Guo X, Wang D, Guo L, Huang T, Dai M (2018) Current status of emerging hypoxia in a eutrophic estuary: the lower reach of the Pearl River Estuary, China. Estuar Coast Shelf Sci 205:58–67. https://doi.org/10.1016/j.ecss.2018.03.004
24. South China Morning Post (2018) How the Greater Bay Area plan can boost Hong Kong's environmental collaboration. Available at https://www.scmp.com/comment/insight-opinion/article/2148240/how-greater-bay-area-plan-can-boost-hong-kongs-environmental. Access on 17 Feb 2019
25. Su J, Dai M, He B, Wang L, Gan J, Guo X, Zhao H, Yu F (2017) Tracing the origin of the oxygen-consuming organic matter in the hypoxic zone in a large eutrophic estuary: the lower reach of the Pearl River Estuary, China. Biogeosciences 14:4085–4099. https://doi.org/10.5194/bg-14-4085-2017
26. Sun X, Xu M, Lau YY, Gauri D (2019) Cruisers' satisfaction with shore experience: an empirical study on A China-Japan itinerary. Ocean Coastal Manage 181:1–10
27. Wang G, Li KX, Xiao Y (2019) Measuring marine environmental efficiency of a cruise shipping company considering corporate social responsibility. Mar Policy 99:140–147. https://doi.org/10.1016/j.marpol.2018.10.028
28. Wild P, Dearing J (2000) Development of and prospects of cruising in Europe. Marit Policy Manage 27(4):315–333

Enhancing Sustainable Development Through Regulatory Means and Market-Oriented Incentives for Waste Management in the GBA

Jingyan Fu and Xuebing Lu

Abstract Garbage sorting is conducive to the development of a quality living circle in the Guangdong, Hong Kong and Macau for creating a world-class bay area composed of sustainable cities. This paper focuses on the status quo of the "garbage siege" problems in the Greater Bay Area (GBA) mainly comprising Guangdong, Hong Kong and Macau and highlights the seriousness of the problems associated with waste management. It analyses the market-oriented operation mechanism for urban domestic waste sorting and processing. Under the assumption of a market-based economy, the information asymmetry between "agent" and "principal" is established and different incentive systems are designed to construct a market-oriented model for the classification of urban domestic wastes. Finally, this paper provides some suggestions based on the experience in Guangdong for improving the operations of the waste industry, including two main mechanisms: long-term incentive through regulatory supervision mechanisms as well as market-oriented green finance and investment mechanism, which can jointly be complemented by an integrated waste management policy for enhancing sustainable development of the GBA.

Keywords "Garbage siege" · Waste management · Garbage sorting · Market-oriented operations

---

J. Fu (✉)
Institute of Resource, Environment and Sustainable Development, College of Economics, Jinan University, Guangzhou, China
e-mail: tfjyan@jnu.edu.cn

X. Lu
China Citibank Guangzhou Branch, Guangzhou, China

J. Fu and A. W. Ng (eds.), *Sustainable Energy and Green Finance for a Low-carbon Economy*, https://doi.org/10.1007/978-3-030-35411-4_15

1 Reasons for the Market-Oriented Operation Mode of Garbage Disposal

1.1 Criticalness of Managing Wastes

According to the World Bank report, as early as in 2004, China became the country with the most waste generated in the world. According to the National Bureau of Statistics of China, from 2005 to 2015, the total amount of urban domestic garbage generated in China increased from 156 million tons to 273 million tons per year, and the growth rate reached 75% in only 10 years. Currently, the national municipal solid waste rate is flat annually. Nearly 10% of the growth rate is increasing. Of the more than 660 large- and medium-sized cities across the country, approximately two-thirds of the cities face the problem of "garbage siege". In 2016, the daily output of municipal solid waste in China exceeded 1.4 million tons; it is estimated that by 2020, the total amount of municipal solid waste in China will reach 324 million tons. According to the "2018 Annual Report on Environmental Pollution Prevention and Control of Solid Wastes in Large and Medium Cities in China" issued by the Ministry of Environmental Protection, the total amount of domestic waste generated in 202 large- and medium-sized cities in China reached nearly 202 million tons in 2017, an increase of more than 13 million over 2016. The average output per ton is nearly 1 million tons, which is 13.52% higher than the 890,900 tons in 2016.

1.2 Introducing Market-Based Incentives for Waste Management

With the continuous increase of urban domestic garbage production, the investment and construction of waste sorting and treatment infrastructure has become the key to solving the current urban waste problem in China. At present, China is in a period of rapid economic development, and the multi-faceted capital demand makes the government's fiscal revenue increasingly less likely to meet the needs of urban domestic waste disposal. According to the 2017 China Green Finance Development Report, in recent years, the government had to invest large amounts of financial resources to dispose of urban domestic garbage. In 2016, the national new investment in urban domestic waste treatment was 12 billion yuan, an increase of 253% compared with 2015. In 2017, the national urban domestic waste treatment volume is expected to be 436 million tons, or 1.195 million tons per day, an increase of 87,000 tons compared per day with 2016. The new waste level needs to be increased by 7 billion yuan, which greatly increases the government's financial pressure. In the case of unbalanced capital supply, the introduction of social capital under the guidance of the government and the comprehensive promotion of market-oriented reform of the urban domestic waste industry are the best solutions at present.

1.3 Promoting a Green and Low-Carbon Lifestyle as an Important Component in Developing the Guangdong, Hong Kong and Macau

1.3.1 Development of Guangdong, Hong Kong and Macau into a Quality Living Circle

Under the background of the construction of the Guangdong, Hong Kong and Macau Bay Area, the government has put forward a new era of development and construction requirements for the environment of Guangdong, Hong Kong and Macau. With the promulgation of the "earl River Delta Reform and Development Plan (2008–2020)" and the "Guangdong-Australia Cooperation Framework Agreement" and other policy agreements, the development of economy, environment and other aspects of Guangdong, Hong Kong and Macau is more closely related. In the "Guangdong, Hong Kong and Macau Bay Area Development Plan" issued in February 2019, it is emphasized that the Guangdong, Hong Kong and Macau should be built into a world-class bay area with green low-carbon production, lifestyle, urban development and operations. The "Special Plan for Building a Quality Living Circle", released in June 2012, proposes a regional development vision and cooperation direction for the construction of a quality living circle in Guangdong, Hong Kong and Macau. The long-standing "garbage siege" phenomenon in Guangdong, Hong Kong and Macau is a major "blocking tiger" in the construction of a quality living circle in Guangdong, Hong Kong and Macau, which has seriously hindered the construction and development of a well-developed and sustainable development of Guangdong, Hong Kong and Macau.

2 Problem of "Garbage Siege" in the Context of the Emerging Greater Bay Area

At present, the phenomenon of a "garbage siege" in Guangzhou, the capital city of Guangdong Province, is becoming increasingly prominent. The unbalanced development of the environment and economy will constrain the steady growth of Guangzhou's economy. The need for garbage disposal will promote the market-oriented reform of the domestic waste industry in Guangzhou. As the capital city of Guangdong Province and the central city of the country, Guangzhou is the main city of the Bay Area of Guangdong, Hong Kong and Macau. With the rapid development of its economy, the amount of domestic garbage generated is also increasing year by year. It is a typical "garbage siege". This finding runs counter to the development planning goals of Guangdong, Hong Kong and Macau. In 2016, the total amount of municipal solid waste generated in China was approximately 204 million tons, of which the amount of domestic garbage produced in Guangdong Province was 23.91 million tons, while the domestic garbage produced in Guangzhou reached 5,352,100 tons, accounting for the national municipal solid waste. The total amount produced was 2.62%, accounting for 22.38% of the total amount of domestic garbage produced in Guangdong Province. As of December 2017, Guangzhou has a resident population of 14.408 million, of which the urban population is 12.48 million and the urbanization rate is 86.14%. Its population size and urbanization process are China's leading representative cities. According to the "Guangzhou City Master Plan (2017–2035)", the permanent population of the city will reach 20 million by 2035. It is foreseeable that a large amount of urban domestic waste will have a serious impact on the current domestic garbage disposal system in Guangzhou. The continuous growth of Guangzhou's population and the increased consumption levels have led to a growing trend in the production of domestic waste. The demand for development funds for the waste industry has been far greater than supply, and the government's fiscal revenue is increasingly difficult to meet the development of the domestic waste industry in Guangzhou. Reality needs. Therefore, the economic vitality of Guangzhou's domestic waste industry can be improved through market-oriented reforms. Waste sorting is a low-efficiency traditional industry with great potential for reform and is an indispensable part of Guangzhou's search for new economic growth points.

Hong Kong and Macau have also faced the problem of "garbage siege" in recent years. According to the Hong Kong Solid Waste Monitoring Report 2017 and the Macau Environmental Status Report 2017, the amount of municipal solid waste and per capita urban solid waste disposed of in Hong Kong and Macau has continued to rise in recent years. The per capita daily solid waste and disposal volume in 2017 were 1.45 kg and 2.16 kg, respectively, which were higher than 0.98 kg in Guangzhou. This finding reflects that the garbage problem in Hong Kong and Macau is also very prominent, which seriously restricts the ecological construction development of Guangdong, Hong Kong and Macau. As the leading soldier in the reform and opening up of the country, and an important bridge connecting Hong Kong and Macau, Guangzhou City should undoubtedly shoulder the heavy responsibility of "pre-testing" in the market-oriented reform of the garbage industry, paying attention to and drawing on the waste treatment dynamics and experience of the two major cities of Hong Kong and Macau. Based on this, we will promote the construction of the Pearl River Delta living circle with a high-quality garbage sorting environment.

With economic development, population growth and the improvement of people's living standards, the phenomenon of "garbage siege" in Guangzhou has increasingly constrained the sustainable development of cities. As the third largest city in China and the capital city of Guangdong Province, Guangzhou is China's southern gateway to the world. It is the main city of the Bay Area of Guangdong, Hong Kong and Macau, the Pan-Pearl River Delta Economic Zone and the political, economic and cultural centre of South China. Since the reform and opening up, Guangzhou has made outstanding achievements in economic and cultural construction and has strong international influence. Generally, the most important factor in all economic development is population resources. Because the more abundant the population resources are, the more participants in the urban economic construction, and the greater the city scale is expanded. However, the economic development and population expansion of large cities, environmental pollution and other issues are more prominent. If they are not effectively resolved, they will inevitably lower the living standards of residents and pose a serious threat to the sustainable development of the urban economy. As an economically developed city in South China, Guangzhou is a rapidly changing city, attracting many migrants to live in this agglomeration. The number of permanent residents in the city is increasing year by year. As of December 2017, the resident population of Guangzhou was 144.884 million, of which the urban population was 12.488 million and the urbanization rate was 86.14%. It is a veritable megacity in China. The various garbage problems brought about by this population growth have become an urgent problem to be solved in the modernization and economic development of Guangzhou.

In terms of absolute quantity, the output of urban domestic garbage in Guangzhou has increased year by year. Figure 1 shows the permanent population and urban domestic waste production in Guangzhou from 2010 to 2017. It can be seen or calculated from the following. First, during the period of 2010–2017, the resident population of Guangzhou increased continuously, while the total amount of municipal solid waste generated by the city grew at an average annual growth rate of more than 10%, far exceeding the growth rate of the resident population in Guangzhou. In 2015, the output growth rate was as high as 51%. Second, the total amount of domestic garbage produced in Guangzhou has shown an overall upward trend. In 2017, the total amount of production of domestic garbage reached 7,737,600 tons, and the daily output reached 20,200 tons. The per capita daily output was as high as 14 tons. If the capacity of an ordinary garbage truck is 10 tons, then the daily garbage produced in Guangzhou is as high as 2000 cars, and the average person produces nearly 1.4 cars of garbage per day. As the living standards and per capita consumption levels of urban residents in Guangzhou are still in the advancement stage, it is expected that future urban domestic garbage production in Guangzhou will continue to rise.

In terms of relative quantity, the output of urban domestic garbage in Guangzhou has always been among the highest in the country. Figure 2 shows the top ten cities in China in 2013–2017. The following findings can be seen or calculated. First, the

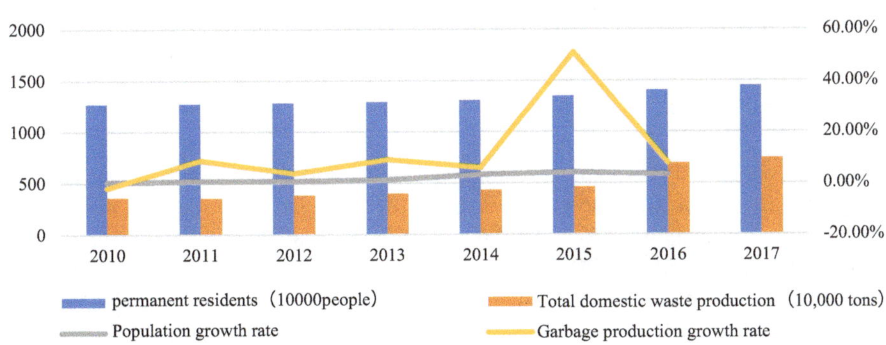

Fig. 1 Permanent population and urban domestic waste production in Guangzhou from 2010 to 2017. *Source* http://www.gzcgw.gov.cn

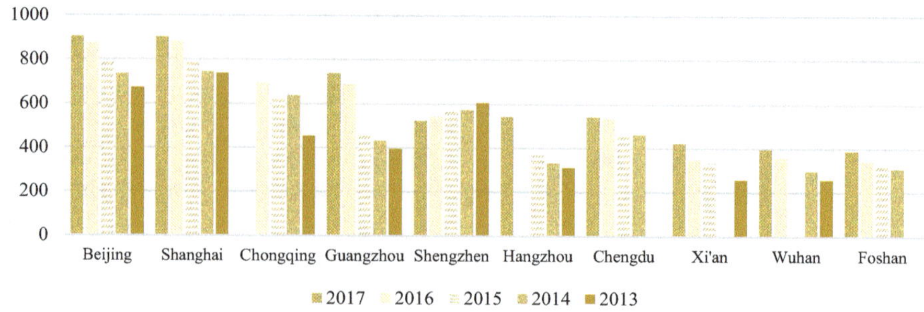

Fig. 2 Top ten cities in China for urban domestic garbage generation in 2013–2017. *Source* http://www.mee.gov.cn

amount of municipal solid waste generated in Guangzhou has been among in the top six in the country in recent years. The amount of domestic garbage generated by large- and medium-sized cities was 202 million tons, of which Guangzhou's output was 737,660 tons; Guangzhou was ranked third, second only to Beijing and Shanghai, accounting for 3.65% of the 202 large- and medium-sized cities nationwide. The phenomenon of garbage siege is prominent.

In general, for a long time, the amount and growth rate of domestic garbage in Guangzhou has far exceeded its limited waste disposal capacity. The contradiction between urban sustainable development and garbage disposal has become increasingly prominent. The phenomenon of "garbage siege" has existed for a long time. The unilateral dependence on government governance and financial supply has long been unable to meet the requirements of social development in terms of capital guarantee and operational efficiency. Further promoting the market-oriented operation of urban domestic waste sorting has become an inevitable requirement for Guangzhou.

3 Literature Review

3.1 Necessity of Market-Oriented Operations of the Waste Industry

Research by Western scholars on the market-oriented operation of the waste industry is mainly carried out from the perspectives of the theory, the comparison before and after the market-oriented reform, and the efficiency; this research is combined with field investigations and interviews for analysis and has accumulated detailed research results. On a theoretical level, Joseph [1], based on the theory of competitive market, and Arena [2], based on the theory of effective competition, noted the necessity of introducing market forces in the process of waste sorting and provided a new connotation for the treatment of urban domestic waste. On a practical level, Sabbas et al. [3] conducted a field survey of 48 small towns in Canada and found that the cost of relying solely on government-sponsored garbage collection was higher than that of contracted garbage collection. The necessity of market-oriented operation of the waste industry. Oteng-Ababio et al. [4] conducted a survey of 101 cities in Africa and found that the average cost of directly operating the garbage industry in the public sector is higher than that of private enterprises. Themelis and Mussche [5] found that the government's garbage collection costs were 124% higher than public–private partnerships by comparing 198 government agencies in the US, Japan and China. Finstein and Morris [6] conducted field surveys and interviews in 33 cities in 1970 and found no significant efficiency differences between government departments and the private sector.

3.2 *Realization Path of the Market Operation of the Garbage Industry*

Western countries have begun to study government public utilities early. Therefore, most of the research on market-oriented operation of the urban domestic waste sorting and processing industry is based on the reform model of government public utilities, which has formed a series of theoretical results. Ferreira et al. [7] analysed in detail the economic factors affecting the production of domestic waste and compared the benefits before and after the implementation of the waste volume charging system. It was proposed that the specific charging system can promote the reduction and market operation of garbage to a certain extent. Okuda and Thomson [8] proposed that the recycling space of urban domestic garbage is large and that space is large. Public participation, the introduction of third-party funds, the construction of classification and treatment facilities are important factors affecting the market operation of domestic waste.

3.3 *Performance on the Market Operations of the Waste Management Industry*

Research by Western scholars on the market-based performance of the government's public sector focuses on the three dimensions: market structure, ownership structure and regulatory reform; the scholars use relevant performance variables to assess market-based reform performance, while government public utilities reform performance. The research has important guiding significance for the market-oriented reform of the waste industry. In empirical research, Igarashi et al. [9] conducted an empirical analysis of garbage collection enterprises, pointing out that private enterprises have an advantage in improving the internal operation efficiency of the waste sorting industry. The empirical findings of Beigl et al. [10] reflect that the cost of the waste industry mainly depends on the competitive external environment and that the public–private property rights after market reforms have merged. Eriksson et al. [11] analysed the impact of both state-owned and private ownership systems and noted that privatization of public goods should be addressed. In terms of market structure, Khetriwal et al. [12] believes that market competition not only optimizes the allocation of resources but also effectively motivates related companies to improve the efficiency of waste sorting. Kikuchi and Gerardo [13] noted that regulatory reform not only promotes market-oriented reforms by improving the efficiency of business operations but also improves the pricing mechanism of public services.

This paper examines Guangzhou as the target city, systematically analyses the development status of all aspects of the market-oriented operation of Guangzhou municipal waste separation and the problems of the three major participants of government, enterprises and residents, and combines the case of enterprise green investment to support waste classification projects. The paper tries to build a market-oriented operation model for the "three-in-one" classification of domestic wastes for Guangzhou, which is tailored to local conditions, and clarify the roles and responsibilities of relevant participants.

4 Analysis of the Market-Based Mechanism of Urban Domestic Garbage Classification and Processing

In the market economy, most of the subjects present a situation where the information is not symmetrical, and this will lead to a principal–agent relationship. If the parties in the market are assumed to be rational economic agents, they pursue the maximization of utility, in which entrepreneurs pursue profit maximization. Under the premise of this assumption, the principal and the agent will perform multiple rounds of game to maximize the expected utility and finally reach a relative equilibrium state. In this process, because the information is not symmetrical for a long time, the principal often uses incentives to effectively stimulate the agent's behaviour, stimulate demand and enhance its internal behavioural motivation to achieve organizational goals. In this paper, the government and other relevant departments do not fully understand the efforts of enterprises and residents to participate in the classification and disposal of municipal solid waste. Therefore, enterprises and residents can be considered "agents" and the government "clients"; enterprises have information asymmetry, which requires that in the process of deepening the garbage revolution, the government needs to formulate an effective incentive system to encourage and control the behaviour of enterprises and residents and to fully stimulate the garbage separation and recycling behaviour of multiple entities.

This paper argues that the incentive mechanism for the classification and treatment of urban domestic waste refers to government or government joint ventures adopting legal means, administrative means, economic means, etc., through incentives, punishments, fees, investment, publicity, pilots and other measures to encourage the public to participate in domestic garbage. Classification processing. Specifically, incentives under incentives should include positive and negative incentives, material rewards and spiritual rewards. Among them, positive incentives refer to the promotion of public waste sorting behaviours, such as prizes and bonuses, through rewards. Negative incentives can also be seen as punishments, which mean that, through sanctions, incentives are imposed on non-compliant waste sorting behaviours, such as fines, notices of criticism and deductions. Material rewards refer to motivating the public to participate in the waste sorting process, such as bonuses and daily necessities, from meeting the material needs of the public. Spiritual reward refers to the purpose of satisfying the public's spiritual needs and influencing the public's psychological state, such as awarding honorary titles. From the perspective of economic means, the incentive system can further specifically divide incentives such as government subsidies, incentives for recycling enterprises, tax incentives and environmental protection points, as well as constraints such as minimum target recovery rate, payment system, deposit system and fee penalties. Large classes. From the specific aspects of waste sorting, the incentive system can be further divided into an incentive system for waste placement and collection, transportation and processing, such as the possibility of increasing the expected profit of the relevant enterprises in garbage discharge and collection, in garbage transportation and the processing links that subsidize related companies.

Figure 3 is a summary of the market-oriented model for the classification of urban domestic waste. Based on the analysis of the above theory, this paper adds the role of market factors and incentive mechanisms in the traditional urban waste separation process. It can be seen that the market-oriented operation from garbage sorting to classified collection, sorting, clearing and end-processing is inseparable from the participation of urban residents, enterprises and government multi-agents and that the

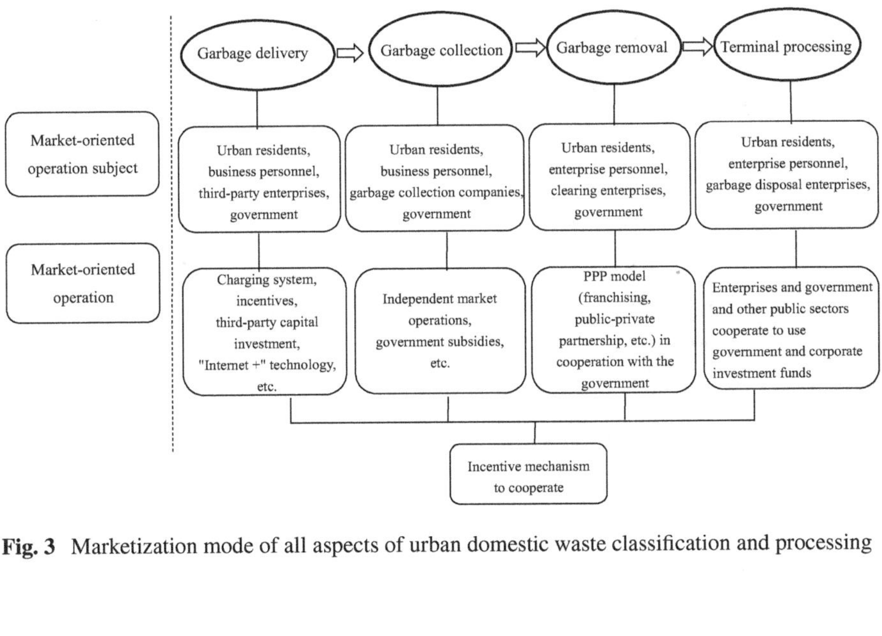

Fig. 3 Marketization mode of all aspects of urban domestic waste classification and processing

higher the degree and investment of third-party enterprises are, the more favourable it is to promote the market-oriented development of the waste industry. As far as the current situation is concerned, it is necessary to clarify the relationship between the government and the market in each link and to solve the problem of domestic waste such as investment funds, waste sorting accuracy and harmless processing capacity through the use of multiple market forces while creating and inspiring more profit points, promoting the economies of scale of third-party enterprises and building a modernized and efficient market-oriented operation system for urban domestic waste sorting.

5 Concluding Notes: Improving the Market-Based Operations of the Waste Management Industry

Under a constant development of the economy, more domestic garbage will further increase resources required for waste management and create burdens on the public funding. To reduce such a liability to the public funding government while fostering sustainable development, the Guangzhou municipal government should try to alter its roles and to formulate policies for regulating the practice of the waste management industry. As authority to the industries, the government is expected to lead and facilitate market-oriented operation mechanism, actively introduce social capital, participating in waste separation and recycling, reduce the input cost of waste separation processing intermediates by strengthening system construction, as well as to improve the efficiency of garbage disposal and promote the development of the waste management industry.

5.1 Formulating a Clear Strategic Goal for the Market-Oriented Operations of the Waste Management Industry

The market-oriented operation of the municipal solid waste industry in Guangzhou, for instance, should clarify two important overall strategic objectives. The first objective is to construct a rational market-oriented investment operation model for the domestic waste industry. We should break the traditional situation of domestic waste sorting and handling by the government and financial operation, vigorously introduce social capital, promote investment diversification and operational marketization in the domestic waste industry and gradually transform the behaviour of the original social welfare undertaking into the government. Market and economic behaviours should guide and supervise the market and operate as an investment entity. The second objective is to fully realize the source reduction of the domestic waste classification link and promote the construction of a waste-free city. At present, the garbage revolution in Guangzhou is at the end of the garbage disposal period in Japan in 1970. Therefore, Guangzhou has taken the end-processing capacity of garbage disposal enterprises as the focus of the market development of the waste industry and to some extent ignored the importance of waste source reduction. In the process of further promoting the market-oriented operation of the domestic waste industry in Guangzhou in the future, we should pay attention to improving the terminal processing capacity of waste disposal enterprises and gradually shifting to reducing the amount of garbage generated from the source, that is, the amount of garbage generated before is an independent variable. The amount of treatment is the dependent variable. The amount of garbage generated will be landfilled or incinerated to treat the corresponding amount of waste. In the future, it will need to be reversed to set the final amount as the ceiling, effectively reducing the landfill and incineration amounts.

Figure 4 shows two distinct rationales for the garbage revolution. At present, Guangzhou mainly uses the rational of the left picture to promote the garbage sorting work, which focuses on improving the harmless end-disposal ability, but this paper believes that the future rational is inevitable. The future rational is in the right picture, which focuses on the reduction of the source, reducing the amount of harmless treatment.

At the same time, on the basis of clarifying the two overall strategic objectives, we should follow the relevant policy trends in China and formulate the phased strategic objectives for the market-oriented operation of the municipal solid waste industry in Guangzhou according to local conditions. Figure 5 is the "three-step" phased strategic goal of Guangzhou City to promote the "garbage revolution". In the first stage, from now until 2020, the Guangzhou municipal government should vigorously introduce social capital to invest in the waste industry. At the same time, it should gradually implement the mandatory classification of garbage through legislation and improve the end-processing capacity of waste, from garbage mixed landfill to classification. In

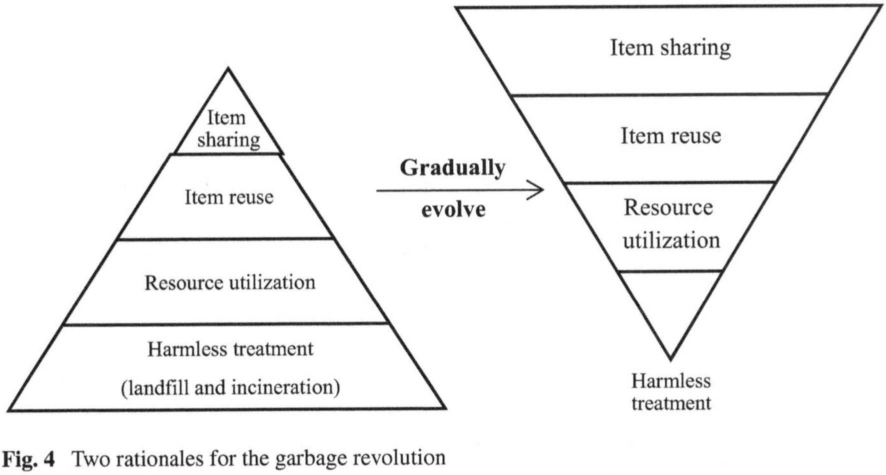

Fig. 4 Two rationales for the garbage revolution

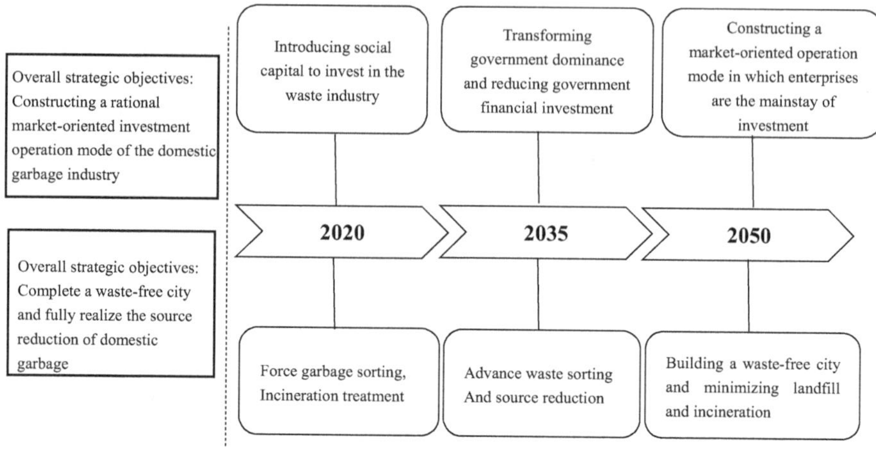

Fig. 5 Conception of phased strategic targets for promoting the "garbage revolution" in Guangzhou

the second stage, from 2020 to 2035, the Guangzhou municipal government should comprehensively change the status of the government-led market-oriented operation management of the waste industry and reduce the government's financial input. At the same time, it must shift from positive growth in waste production to negative growth. In an economically low-waste society, the amount of landfill and incineration will be reduced, and waste sorting and source reduction will be realized. In the third stage, from 2035 to 2050, a market-oriented operation mode of the domestic waste industry will be constructed, which is operated by enterprises. The goal of building a city without waste is to minimize landfill and incineration amounts.

5.2 *Formulating and Implementing Long-Term Incentives and Regulatory Supervision Mechanisms to Deepen the Market-Oriented Development of the Waste Management Industry*

Compared with the traditional regulatory restraint system, the development of a long-term effective incentive mechanism can motivate Guangzhou residents to participate in garbage sorting and can produce better results. Seacat and Northrup [14] noted that "it is more important for the government not to care about whether the public understand how to classify waste, but to motivate the public to act for the benefit of society". Japan's strict penalties and regulatory mechanisms and a diverse support and incentive system are important aspects for the success of its waste sorting process. As a guide to promote the market-oriented operation of the domestic waste industry in Guangzhou, the Guangzhou municipal government should formulate and implement long-term effective economic incentives. On the one hand, the government should actively guide more social capital to invest in the waste industry; on the other hand, the government should fully mobilize Guangzhou. City residents will be motivated to classify waste and deepen the market-oriented development of the waste industry.

According to different incentive methods and incentive subjects, economic incentives can be divided into three categories: taxation, subsidy and a fee system. Taxation and charging systems are the main economic means. In terms of the charging policy, it can be applied to different targets through economic incentives such as volume charging, product charging and a landfill tax. Specific charging mainly promotes garbage recycling and source reduction and alleviates government funding pressure. Product charges are used to deprive products that produce "high waste" from the source of production. The landfill tax is to promote the transformation of waste disposal methods and resource recycling. In terms of subsidy policy, it can be applied to different objects through economic incentives such as subsidies, tax incentives, a reduction or exemption, government priority procurements, recycling credits and primary resource taxes. In addition, a deposit refund system can be adopted to curb consumption, stimulate corporate recycling, promote consumer return and recyclable products, organize garbage collection activities and set up environmental sanitation awards to stimulate public waste sorting.

At the same time, with reference to the Japanese experience, while establishing a long-term incentive mechanism, synergy with market-based regulatory mechanisms will be more effective. In terms of supervision, the participation of professional third parties, the media, the National People's Congress, the CPPCC, the public and other multi-supervised entities, use inspection, processing methods, spot checks, inspections, audits, fees, etc., can be used to reduce the amount of garbage generated. Supervision, such as investment company's access and exit supervision, related facilities construction and operation supervision, garbage pollution and garbage disposal secondary pollution supervision, regulations and policies, policy formulation, implementation and feedback supervision, etc., severely punish enterprises or residents for illegal investment operations. Acts such as stealing, smuggling, wasting resources, polluting the environment, dereliction of duty, and standardizing and supervising the participants in the waste sorting and processing industry from the two levels of entity and procedure ensure the market-oriented development of the garbage industry in Guangzhou.

5.3 *Building a Market-Oriented Investment Mechanism for the Domestic Waste Industry and Regulating the Investment Market for the Waste Management Industry*

From the perspective of marketization, the sources of funding for the waste management industry should be composed of both public and private funding. The proportion of the waste management industry is a dynamic process that is affected by factors such as economic development, government macro-control and market operation efficiency. At a time when the market-oriented investment operation of the domestic garbage industry in Guangzhou is in relative infancy, the Guangzhou municipal government should gradually build a mechanism to promote enterprises to invest in the domestic waste industry, vigorously introduce social capital and give full support to the superiority of the market mechanism for enterprises. Investors should carry out investment operations in the waste management industry to create a good development environment. For example, by leveraging on the green finance and investment trends, we can encourage and guide third-party environmental protection companies to develop relevant markets, expand the scope of waste separation, promote the transformation and upgrading of the waste industry structure, and promote investment cooperation and common use in emerging industries such as the domestic waste industry and the Internet development.

5.4 Improving Domestic Garbage Disposal Charging Policy and Market-Oriented Dynamics of the Waste Management Industry

On the one hand, the domestic garbage disposal charging policy is the economic means that best reflects the will of the government. The domestic garbage disposal cost is also direct and rapid, which can effectively reflect the government's management orientation towards the market-oriented operation of the waste industry. On the other hand, the cost of domestic garbage is an important guarantee for the market-oriented operation of the waste industry, and it is a profit point that investment companies are very concerned about in the future, which can promote the problem of solving the bottleneck of garbage management. From the perspective of externality theory, garbage collection is reasonable. It not only combines personal interests with garbage classification but also breaks the masses' dependence on the government's inertia thinking in garbage classification and can effectively promote the source reduction of garbage. In July 2018, the National Development and Reform Commission issued the "Opinions on Innovation and Improvement of the Green Development Price Mechanism", proposing that by the end of 2020, national cities should comprehensively establish a domestic garbage disposal fee system.

For instance, Guangzhou City implements a household-based charging system, which has problems such as low awareness by residents and poor results in reducing the source of garbage. Drawing on the experience of Japan, according to the actual situation in Guangzhou, we can consider implementing a differential metering and charging system on the principle of "pay more for garbage, pay less and waste less". First, for the garbage delivery, the weight of garbage or the volume is calculated for the corresponding garbage disposal fee. Second, for the garbage collection, additional recycling and disposal costs should be charged for large pieces of garbage, such as furniture and electrical appliances. The effect of the domestic garbage disposal charging system on the classified behaviour of Guangzhou residents is mainly reflected in its economic incentive and binding mechanism. If the metering and charging policy is implemented, the residents should pay more for garbage disposal. Through economic means, it restricts the garbage classification behaviour of residents, and at the same time, it can also stimulate the behaviour of residents who have less garbage. Due to the consideration of economic factors, residents are driven to reduce the generation of domestic garbage to a certain extent and consciously carry out the classification of garbage, which will constrain the behaviour of residents. Such policy for waste management can be introduced to the GBA so as to develop a shared value system as well as an integrated approach that will enhance sustainable development in the region.

References

1. Joseph K (2006) Stakeholder participation for sustainable waste management. Waste Manage 30:863–871
2. Arena U (2012) Process and technological aspects of municipal solid waste gasification. A review. Waste Manage 4(32):625–639
3. Sabbas T, Polettini A, Pomi R et al (2003) Management of municipal solid waste incineration residues. Waste Manage 9(23):61–88
4. Oteng-Ababio M, Arguello JEM, Gabbay O (2013) Solid waste management in African cities: sorting the facts from the fads in Accra, Ghana. Habitat Int 39(39):96–104
5. Themelis NJ, Mussche C (2001) Municipal solid waste management and waste-to-energy in the United States, China and Japan. J Environ Manage 62:19–28
6. Finstein MS, Morris ML (1975) Microbiology of municipal solid waste composting. Adv Appl Microbiol 6(19):113–151
7. Ferreira C, Ribeiro A, Ottosen L (2003) Possible applications for municipal solid waste fly ash. J Hazard Mater 1(96):201–216
8. Okuda I, Thomson VE (2007) Regionalization of municipal solid waste management in Japan: balancing the proximity principle with economic efficiency. Environ Manage 7(40):12–19
9. Igarashi M, Hayafune Y, Sugamiya R et al (1984) pyrolysis of municipal solid waste in Japan. Energ Res 106(3):377–382
10. Beigl P, Lebersorger S, Salhofer Stefan (2008) Modelling municipal solid waste generation: a review. Waste Manage 5(28):200–214
11. Eriksson O, Carlsson Reich M, Frostell B et al (2005) Municipal solid waste management from a systems perspective. J Clean Prod 2(13):241–252
12. Khetriwal DS, Kraeuchi P, Widmer R (2009) Producer responsibility for e-waste management: key issues for consideration-learning from the Swiss experience. J Environ Manage 90(1):153–165
13. Kikuchi R, Gerardo R (2009) More than a decade of conflict between hazardous waste management and public resistance: a case study of NIMBY Syndrome in Souselas. J Hazard Mater 172:1681–1685
14. Seacat JD, Northrup D (2010) An information–motivation–behavioral skills assessment of curbside recycling behavior. J Environ Psychol 30(4):393–401

Kolibri 360 GmbH
Ferdinand-Jühlke-Straße 7
99095 Erfurt, Deutschland
produktsicherheit@kolibri360.de